Advances in Data Clustering

Fadi Dornaika • Denis Hamad •
Joseph Constantin • Vinh Truong Hoang
Editors

Advances in Data Clustering

Theory and Applications

Springer

Editors
Fadi Dornaika
Department of Computer Science
and Artificial Intelligence
University of the Basque Country
San Sebastián-Donostia, Guipúzcoa, Spain

Joseph Constantin
LaRRIS, Faculty of Sciences 2
Lebanese University
Fanar, Lebanon

Denis Hamad
University of the Littoral Opal Coast
Calais, France

Vinh Truong Hoang
Department of Information Technology
Ho Chi Minh City Open University
Ho Chi Minh City, Vietnam

ISBN 978-981-97-7678-8 ISBN 978-981-97-7679-5 (eBook)
https://doi.org/10.1007/978-981-97-7679-5

This Springer imprint is published by the registered company Springer Nature Singapore Pte Ltd.
The registered company address is: 152 Beach Road, #21-01/04 Gateway East, Singapore 189721, Singapore

Preface

Clustering, a foundational technique in data analytics, holds diverse applications across scientific, technical, and business domains. This book, themed around "Data Clustering," assumes substantial importance due to the indispensable role clustering plays in various contexts.

In the current era of online media and digital communication, large datasets are generated at an unprecedented rate. This proliferation of data underscores the importance of clustering as a pivotal technique in data mining and machine learning. At its core, clustering aims to identify heterogeneous groups within unlabeled data, making it a crucial unsupervised task in machine learning. The primary objective of clustering is to automatically assign meaningful labels to each unlabeled datum with minimal human intervention. By analyzing this data, we can categorize information, uncover patterns, and draw actionable conclusions that are applicable across a wide range of application domains, from healthcare and finance to marketing and social media analysis.

The main challenge with unlabeled data is defining a quantifiable goal to guide the model-building process, which is the central theme of clustering. Unlike supervised learning, where the presence of labeled data provides a clear objective, unsupervised learning through clustering must derive its objectives from the inherent structure of the data itself. This requires sophisticated algorithms capable of discerning the underlying patterns and relationships within the data without prior knowledge or labels.

Over the past decades, numerous clustering methods based on shallow models have been proposed. These traditional approaches include techniques such as k-means, hierarchical clustering, and Gaussian mixture models, which have been widely used due to their simplicity and effectiveness in various scenarios. However, these methods often fall short when dealing with complex, high-dimensional data that is common in modern applications. The advent of deep neural networks has revolutionized the field of machine learning, and researchers have increasingly explored paradigms that harness the power of deep learning for clustering tasks. Deep clustering leverages the hierarchical feature extraction capabilities of deep neural networks to improve clustering performance. This approach is applicable to both non-graph data, such as image collections, where deep convolutional networks can capture intricate visual patterns, and graph-based structures, like social networks, where graph neural networks can model the relationships between entities. The driving force behind the adoption of deep clustering is the recognition that deep representations excel at extracting highly valuable features from raw data. These features, which are often difficult to capture with shallow models, enhance the ability of clustering algorithms to group similar data points accurately and uncover more meaningful structures within the data.

Moreover, deep clustering methods often integrate advanced techniques such as autoencoders, variational inference, and reinforcement learning to further refine the clustering process. Autoencoders, for instance, can learn compact and informative representations of the data, which are then used for clustering. Variational inference provides a probabilistic framework that helps in managing uncertainty and capturing complex data distributions. Reinforcement learning can guide the clustering process by optimizing specific clustering objectives through trial and error.

In summary, this book delves into the evolution and advancements in clustering techniques, emphasizing the transition from traditional shallow models to cutting-edge deep learning approaches. It explores the theoretical foundations, practical implementations, and diverse applications of clustering in modern data analytics. By leveraging the power of deep neural networks, we can significantly enhance clustering performance, thereby enabling more effective and insightful analysis of large and complex datasets.

San Sebastián-Donostia, Spain — Fadi Dornaika
Calais, France — Denis Hamad
Fanar, Lebanon — Joseph Constantin
Ho Chi Minh City, Vietnam — Vinh Truong Hoang

Acknowledgment

First and foremost, we extend our deepest gratitude to all the contributing authors who have dedicated their time, effort, and expertise to make this compilation a success. Your insightful chapters form the cornerstone of this book, and without your contributions, this work would not have been possible.

We would also like to thank our publisher, Springer, for believing in this project and providing the resources and platform to bring it to fruition. Special thanks to the editorial and production teams for their hard work and professionalism. A heartfelt thanks to our colleagues and peers who provided feedback and encouragement during the development of this book. Your constructive critiques and suggestions have significantly enriched the final product.

Finally, we extend our appreciation to all the readers who will engage with this book. It is our sincere hope that the chapters within will inspire, inform, and contribute meaningfully to your work and knowledge.

Editors,

San Sebastián-Donostia, Spain — Fadi Dornaika
Calais, France — Denis Hamad
Fanar, Lebanon — Joseph Constantin
Ho Chi Minh City, Vietnam — Vinh Truong Hoang

Contents

Chapter 1
Classification of Gougerot-Sjögren Syndrome Based on Artificial Intelligence

A. Olivier, A. Mansour, C. Hoffmann, L. Bressollette, S. Jousse-Joulin, and B. Clement

Abstract Gougerot-Sjögren syndrome (GSS) is an incurable chronic autoimmune disease that involves an inflammatory process and lymphoproliferation that primarily affects the lacrimal and salivary glands. This disease mainly affects women (the ratio of affected women can be nine times higher than the ratio of affected men). According to an epidemiology study, GSS at different severity levels may affect between 0.1 and 5% of the total population. Usually, GSS detection is performed by biopsy. Some medical studies showed a correlation between biopsy results and the salivary gland ultrasonography (SGUS). On the other side, ultrasound imaging devices are widely used in various medical fields thanks to their noninvasive nature, safety and nonimpact on patients' health. However, these grey images are affected by noise and artifacts. In our project, we developed an artificial intelligence approach to classify and detect GSS only based ultrasound imaging. Indeed, the salivary glands are made of tissue, with acinar, ductal, and myoepithelial cells. Some sonographic features are clearly identified for the detection of the primary GSS. Additionally, some patterns in the textures can help differentiate GSS with other diseases. So, we extracted specific features and then developed a learning scheme for deep neural networks based on joint training on classification and segmentation tasks. We obtained conclusive accuracy on the detection of GSS.

Keywords Deep learning · Data fusion · Pulmonary embolism · Ultrasound imaging

---

A. Olivier · A. Mansour (✉)
ENSTA, Lab-STICC UMR 6285 CNRS, Brest, France
e-mail: aurelien.olivier@ensta-bretagne.org; mansour@ieee.org

C. Hoffmann · L. Bressollette · S. Jousse-Joulin
GETBO UMR 13-04 CHRU Cavale Blanche, Brest, France
e-mail: luc.bressollette@wanadoo.fr; sandrine.jousse-joulin@chu-brest.fr

B. Clement
ENSTA, Lab-STICC UMR 6285 CNRS, Brest, France

CROSSING IRL CNRS, Adelaide, SA, Australia
e-mail: benoit.clement@ensta-bretagne.fr

F. Dornaika et al. (eds.), *Advances in Data Clustering*,
https://doi.org/10.1007/978-981-97-7679-5_1

1.1 Introduction

The Gougerot-Sjögren syndrome (GSS) is an autoimmune disease that involves an inflammatory process and lymphoproliferation that primarily affects the lacrimal and salivary glands. Common symptoms include xerostomia (dry mouth), keratoconjunctivitis sicca (dry eyes), and enlargement of the parotid gland [17]. This disease affects about 0.5–4.8% of the population [3] and specifically females; indeed, more than 90% of patients affected by GSS are females [3, 11].

The primary GSS concerns patients who have only been affected by GSS; if they suffer from another autoimmune disease, then it is called the secondary GSS. Unfortunately, actual medical treatment can only relieve the disorder symptoms. The challenges in the diagnosis of GSS are related to the fact that its symptoms can be confused with systemic diseases, such as sarcoidosis, amyloidosis, IgG4-related disease, human immunodeficiency virus (HIV), and lymphoma. In fact, these systemic diseases affect the salivary and lacrimal glands and can cause syndromes similar to the ones caused by GSS [14, 25].

The salivary glands facilitate the production of saliva, mastication, swallowing, speech, and taste perception. There are three main salivary glands and 600–1000 small minor salivary glands that can be present throughout the mouth [7]. The three main salivary glands exist on both sides of the face: the sublingual gland is the smallest and closest to the mouth, the submandibular gland, and the largest one is the parotid gland located near the ear.

Usually, GSS detection is performed by biopsy. However, Cornec et al. [4, 21] showed a strong correlation between biopsy results and the salivary gland ultrasonography (SGUS). The latter study motivated several researchers to investigate further noninvasive diagnostic approaches using SGUS. Ultrasound imaging is widely used in various fields such as cardiology, urology, gynecology, and vascular imaging. It is less expensive than other modalities. Additionally, it is fast, noninvasive, and does not require ionizing radiation such as radiography. The drawbacks of ultrasound imaging include signal degradation, resulting in artifacts, noise, and speckles in the images. In our project, we are aiming to improve the automatic detection and segmentation of GSS on SGUS using machine learning and deep learning methods.

Several studies have been published on the segmentation and classification of GSS using ultrasound imaging. Two major approaches have been used for an automatic classification, either training a deep neural network [19] or using feature extraction methods combined with a machine learning classifier [2]. In our previous project [1], we proposed an approach using a scattering operator as a feature extractor to characterize GSS, whereas other works focus on using features based on gray-level textures and statistics, following radiomic approaches [20, 32, 36].

To segment the glands on ultrasound imaging with deep neural networks, Vukicevic et al. [33] compare various architectures of deep convolutionnal neural network (CNN), such as U-Net and FCDenseNet, for the segmentation of salivary glands on 1184 ultrasound images from 287 patients, all diagnosed with primary GSS and reach a dice score of 0.91 with FCN8 (i.e., eight-layer fully convolutionnal neural network). To classify GSS into four classes (definitely GSS, probable GSS, probable non-GSS, and definitely not GSS) using ultrasound images, the authors of [19] apply a VGG16 network pretrained on ImageNet [6]; experts evaluate their confidence in their classification. In [19], the authors used a database containing 200 acquisitions to train their model and obtained an area under curve (AUC)[1] of 0.810 and parotid glands and an AUC of 0.894 on submandibular glands.

In this chapter, we develop classification methods for GSS. Two different ultrasound image databases of salivary glands were used to assess the performance of our approaches. The first database, called GSID, for Gougerot-Sjögren identification, was provided by the Brest University Hospital Center (Brest UHC). The patients included in this retrospective study were suspected to have a GSS. The database contains 210 ultrasound images of normal and ill salivary glands. For each patient, ultrasound images were provided for left and right submandibular and parotid glands. The second database, called HarmonicSS, was gathered in a European project aimed to harmonize the analysis of GSS. We used the GSID database to develop and test our methods. Then, we used HarmonicSS to assess the robustness of our methods. HarmonicSS contains images collected using different devices from various hospitals in several countries. Additionally, the information provided in HarmonicSS is based on various scores related to a gradation of the disease's severity. However, GSID does not contain these scores and only evaluates the occurrence of GSS. Thus, we evaluate an approach to adapt our developed models to the scores introduced in HarmonicSS. We should highlight that a segmentation is required to extract the handcrafted features. These segmentations are obtained through a deep learning model. Then, we analyze our obtained results, using deep learning models for the classification of GSS and for the segmentation of the salivary glands.

1.2 Segmentation Process in GSID

In Gougerot-Sjögren identification (GSID) database, several images are annotated by experts with dot points, while others have binary mask segmentations. For images annotated with dot points, we developed an algorithm in order to extract a binary

[1] The area under curve (AUC) corresponds to the area under the curve of the receiver operating characteristic curve, which plots the true positive rate against false positive rate for various classification thresholds.

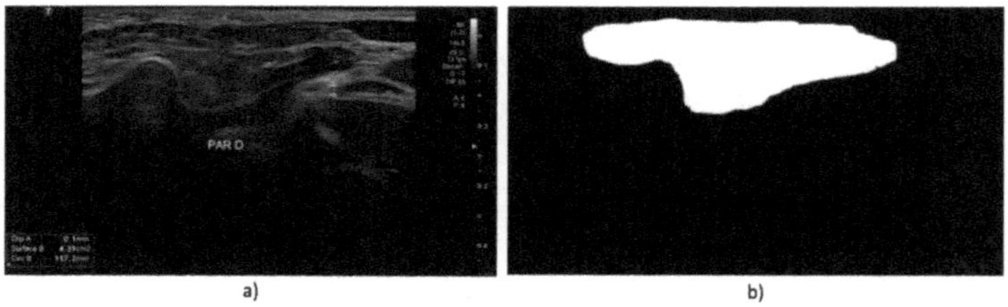

Fig. 1.1 Ultrasound image of a right parotid salivary gland highlighted with a colored dot point contour, and its corresponding automatically extracted binary mask. (**a**) Ultrasound image of a right parotid salivary gland with a colored dot point contour. (**b**) Binary mask providing a segmentation of the salivary gland from the dot point contour in image (**a**)

mask from these annotations; see Fig. 1.1. Our segmentation algorithm uses the following operations:

- **Color threshold:** Selects gray-level values between 200 and 255.
- **Dilation:** The dilation is a morphological operation based on a structural element given by a binary matrix [29].
- **Erosion:** It is also a morphological operation based on a structural element given by a binary matrix.
- **Gaussian blur:** It consists in the convolution of a Gaussian filter with the image. In our project, we applied 3×3 filters to reduce noise and high frequency in the image.
- **Filling holes:** It assigns all pixels within a close contour to the class of the contour.
- **Red-green-blue) (RGB) color map to hue-saturation-value (HSV):** The main advantage of the HSV color model is that it separates the representation of color from its intensity.

For all grayscale images with dot point contours, we used the following operations: The white threshold creates a binary mask. The obtained image still contains other objects than the dot points contour. A Gaussian blur operation reduces the noise in the image. Then, we perform a detection of all isolated components in the image and remove the largest components. This step keeps only the dot points of the contour that have a surface of only one pixel or a few pixels. Then, a dilation operation connects the dots, to remove noisy points and create a thick contour. The two final operations (Gaussian blur and erosion) refine the mask. The erosion reduces the surface of the mask that was over-segmented with the dilation operation, and the Gaussian blur smooths the borders of the segmentation; see Fig. 1.2.

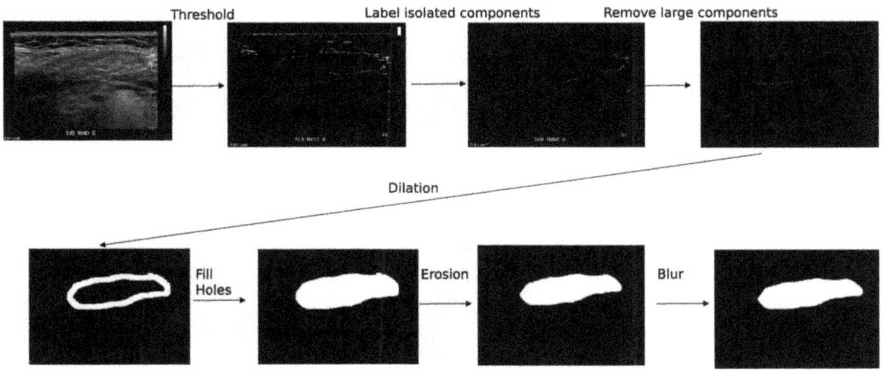

Fig. 1.2 Creation of a binary mask from a white dot points contour

1.3 HarmonicSS Database

HarmonicSS[2] [35] contains 225 ultrasound images of 225 patients from 4 European centers [9]. HarmonicSS proposes several scores grading the severity of the disease:

- The DeVita's score [5] ranges from 0 to 3 in each gland, from normal-appearing morphology to severe inhomogeneity.
- The Outcome Measures in Rheumatology Clinical Trials (OMERACT) score [15] proposes a four-grade classification.
- The European League against Rheumatism (EULAR) classification [28] score is based on five items:
  - Focal lymphocytic sialadenitis[3] and focus score (FS) $\geq$1 by labial salivary glands (LSG)
  - Presence of anti-Ro antibodies, also known as anti–Sjögren's-syndrome (or anti-SSA) antibodies, which are associated with Sjögren syndrome
  - Ocular staining score (OSS $\geq$ 5)
  - Positive Schirmer's test ($\leq$5 mm/5 min): This test determines if the produced quantity of tear is enough to moist the eye
  - Unstimulated whole salivary (UWS) flow rate $\leq$ 0.1 mL/min

Figure 1.3 presents ultrasound images extracted from different centers present in this database.

[2] HarmonicSS refers to **Harmoni**zation and integrative analysis of regional, national, and international **c**ohorts on primary **S**jögren's **s**yndrome (pSS) toward improved stratification, treatment, and health policymaking.

[3] The Focal lymphocytic sialadenitis is defined as the presence of more than 50 lymphocytes around the blood vessels or ducts of the salivary glands.

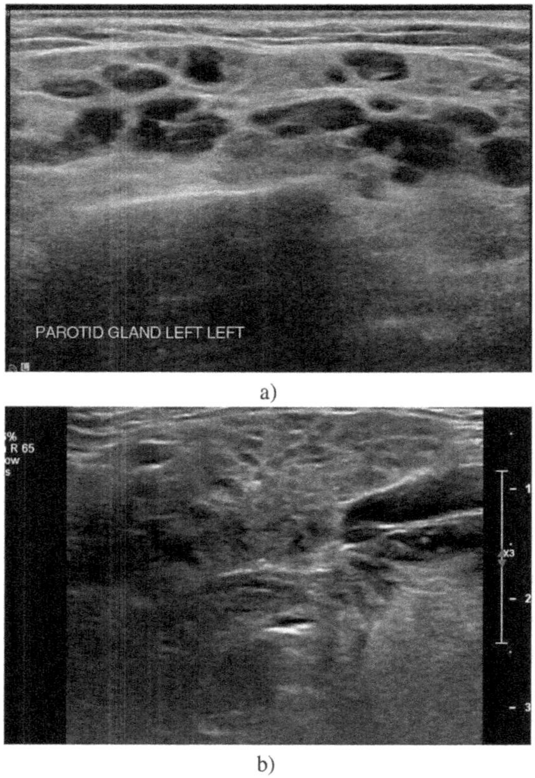

Fig. 1.3 Ultrasound images extracted from HarmonicSS database. (**a**) Acquisition from Udine center with a Samsung device. (**b**) Acquisition from Lubjana center with a Philips device

1.4 Data Pre-processing

In our project, we consider three types of image normalization: raw image without normalization, with a standard normalization (Std) or with a minmax normalization. The standard normalization consists in setting the mean of the image to 0 and its standard deviation to 1. For every pixel value $x(i, j)$ in an image X of size $n \times m$, $i \in [1, n]$ and $j \in [1, m]$, μ is the mean of pixel values within the image X and σ is the standard deviation; the standard normalization is applied as follows:

$$x_{Std}(i, j) = \frac{x(i, j) - \sigma}{\mu} \tag{1.1}$$

The minmax normalization consists in scaling the data to the range [0, 1], by adjusting the pixel value as follows:

$$x_{\min\max}(i, j) = \frac{x(i, j) - \min(x(i, j))}{\max(x) - \min(x)} \tag{1.2}$$

By applying various image transformations [16], the data augmentation is commonly used in order to tackle the lack of data. The transformations are applied to the image consistently with the labels and they can be left and right flip, cropping, sharpening, affine transformation, linear contrast, Gaussian blur, additive Gaussian noise, edge detection, dropout, and elastic transformation. To determine the number of needed transformations, we conduct several experiments modifying that number from 1 to 8. According to our experimental results, we performed inline three augmentations during the training, and we randomly selected three augmentations for each batch.[4]

1.5 Handcrafted Features: Normalization and Selection

This section presents the experiments performed on the classification of the Gougerot-Sjögren syndrome (GSS) using handcrafted features and machine learning. To evaluate the performance, we used the mean decrease impurity (MDI) and minimum redundancy maximum relevance (MRMR).

To normalize gray images based on histogram and following the approaches proposed in PyRadiomics,[5] we used the bin-width method and tested grayscale range width values of 10, 25, and 50 to reach a bin-count from 30 to 130 bins. By performing the bin-width selection with a random forest (RF) classifier and a principal component analysis (PCA) normalization, we obtained a best accuracy of 0.79 with a sensitivity of 0.84 and a specificity of 0.78 with a bin-width of 3. Table 1.1 presents the obtained results for the selection of the best pixel normalization and bin-width on every metric, on average, with a cross-validation.

Table 1.2 shows a comparison of tenfold classification results on all features with a fixed bin-width and different classifiers. The three best results are obtained with either a random forest (RF) or a support vector machine (SVM), with a standard normalization or a PCA.

Testing various feature selections showed better results on the mean accuracy over all classifiers with the features selected by random forest (MDI), as well as the

Table 1.1 Various bin-widths with an RF classifier and PCA

Accuracy	Sensitivity	Specificity	Bin-width
0.79	0.84	0.78	3
0.78	0.82	0.78	5
0.76	0.84	0.73	10
0.74	0.80	0.71	25

[4] A batch is a group of images used in an optimization iteration during the training of a deep learning model.

[5] An open-source python package for the extraction of radiomic features from medical imaging. https://pyradiomics.readthedocs.io/en/latest/.

Table 1.2 Classifier and normalization comparison on a fixed bin-width

Classifier	Feature normalization	Accuracy	Sensitivity	Specificity
RF	PCA	0.78	0.82	0.77
RF	Std	0.80	0.85	0.76
SVM	Std	0.77	0.74	0.82

Table 1.3 Accuracy (Acc) results obtained on all combinations of classifiers, a fixed normalization, and various feature selection methods

Feature selection	Image norm	Dim	Acc mean	Acc max	Acc Std
All features	Std	128	0.70	0.80	0.08
10 selected with MRMR	Std	128	0.72	0.80	0.08
23 selected with MDI	Std	128	0.72	0.84	0.08
23 selected with MRMR	Std	128	0.71	0.79	0.07

maximum accuracy obtained with a single classifier (see Table 1.3). The MRMR-selected parameters provided a slightly lower mean accuracy over all classifiers and a slight decrease for the best accuracy across classifiers compared to MDI. Both feature selections provided an improvement on mean and max accuracy compared to the classifiers trained using all features. Additionally, using the ten most important features selected with MRMR provided better results than using 23. The group of ten features selected with MRMR contained three gray-level co-occurrence matrix (GLCM) features [10], two first-order statistical descriptors, one gray-level run length matrix (GLRLM) descriptor [34], one gray-level dependence matrix (GLDM) feature [30], and two gray-level size zone matrix (GLSZM) features [31]. The 23 features selected with the MDI include 8 GLCM features, 3 GLRLM features, 3 GLSZM features, and 6 first-order features. We find five features in common with the ten MRMR selected features: the cluster tendency in GLCM, the gray-level variance in GLSZM, the mean absolute deviation, and the variance in first-order features.

In this section, we compared the results obtained on the detection of GSS with various classifiers, data normalization, and bin-widths. This work allowed us to find the best settings as a basis for a comparison with the results of deep neural networks. We should highlight that the handcrafted features are extracted from a region of interest (the salivary glands). While we worked on producing segmentation models for the classification of the GSS, our final approach for assessing the potential of handcrafted feature extraction in this work was to consider the annotations to generate the region of interest (ROI) mask. This hypothesis assumes that expert annotations are always provided. However, we also test the training of a deep neural network to perform the segmentation.

1.6 Deep Neural Network on GSS Detection

We used deep neural networks to achieve two primary purposes: segmentation of salivary glands for computing handcrafted features and classification of GSS. We propose here an innovative multiphase training scheme for classification and segmentation tasks. After weight initialization, we introduce the hyperparameters selected with previous experiments. We should highlight that the training is time-consuming and the possibilities of the test of hyperparameters are thus limited. Given these hyperparameters, we present a comparison of several approaches to the multiphase, multitask training on the segmentation and classification. Finally, we evaluate the model on an unseen database with images acquired in different countries and hospitals with several devices.

1.6.1 Weight Initialization

The training of a neural network consists in solving a non-convex optimization problem parameterized over the weights of the network using a backpropagation method [27]. The best weight initialization should set the neural model close to the global minimum. Random initialization with small values of the parameters is commonly used for neural networks, but may not be sufficient if the optimal parameters are large, thus learning process can be slow or even fails to converge. In [13], they experimentally show that gradient learning procedures may not compensate for bad initial values.

For linear activations, the Xavier initialization proposed in [8] initializes the weights using a uniform distribution in $[-\frac{1}{\sqrt{n}}, \frac{1}{\sqrt{n}}]$, where n is the number of neurons in the layer. The authors then empirically proposed to initialize the weights of the ith layer using a uniform distribution in $[-\frac{\sqrt{6}}{\sqrt{n_i+n_{i+1}}}, \frac{\sqrt{6}}{\sqrt{n_i+n_{i+1}}}]$; where n_i represents the number of neurons in the ith layer.

In our project, we used the initialization proposed in [12] which replaces the previous uniform distribution with a centered Gaussian distribution with a standard deviation set for the ith layer, $\sigma_i = \sqrt{\frac{2}{n_i}}$.

1.6.2 Segmentation

An accurate automatic segmentation method can be used to extract features and helps clinicians in their practices. Inspired by the U-Net [26], our model is structured around convolutional blocks. These blocks incorporate convolutional filters, activation functions, and normalization techniques. The model, presented in Fig. 1.4, is divided into two paths:

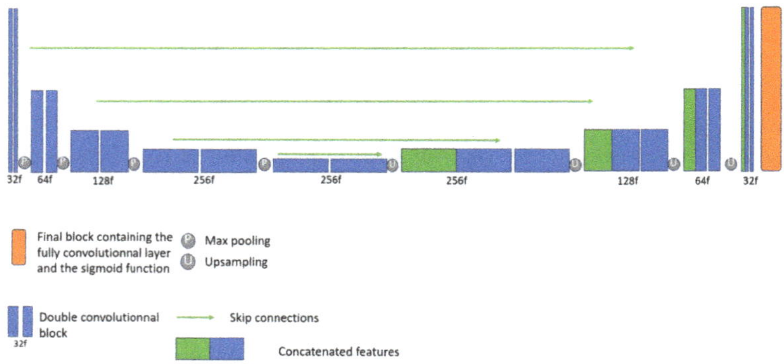

Fig. 1.4 U-Net model adapted

- The encoding path (or down-sampling path) encodes the image into features at various image resolutions. The encoding path uses five double convolutional blocks[6] and four max-pooling operators to generate feature maps,[7] and then down-sample them by a max-pooling operator with a kernel of size 2×2 and a stride of 2. This path finally produces features with dimensions equal to the original image size divided by 16.
- The decoding path (or up-sampling path) reconstructs a segmentation at the original image size from the output of the encoding path. In the decoding path, four double convolutional blocks are used and bilinear interpolation operators between the blocks to up-sample the features.

To expand the receptive field without losing resolution, we use dilated convolutions in the second layer of each double convolutional block as shown in Fig. 1.5. The dilated convolutional uses a kernel of size 3×3 and a stride of 1. The main particularity of the U-Net model lies in the introduction of the skip connections, which involve concatenating feature maps generated in the encoding path with feature maps produced in the decoding path. This concatenation, denoted by the green arrow in Fig. 1.4, is specifically applied to features of the same resolution from both paths. In practice, the output of the first pair of blocks in the encoding path is concatenated with the input of the last block of the decoding path. Following the decoding path, a fully convolutional layer is applied. This layer involves the convolution of two kernels with a size of $1 \times 1 \times C$, where C represents the

[6] A classic double convolutional block (see Fig. 1.5) consists of blocks formed by a convolutional layer with a kernel of size 3×3, followed by a batch normalization and an activation based on a ReLU function [23]. The batch normalization acts on the activations within the network to enhance its convergence. ReLU is chosen for its computational efficiency and its ability to introduce nonlinearity, promoting smooth optimization in the network.

[7] Feature maps refer to the output of layers within the model. They represent modified versions of the original images that capture specific attributes, such as edges or gradients in the case of the first layer. As we progress through the layers, the features become more complex.

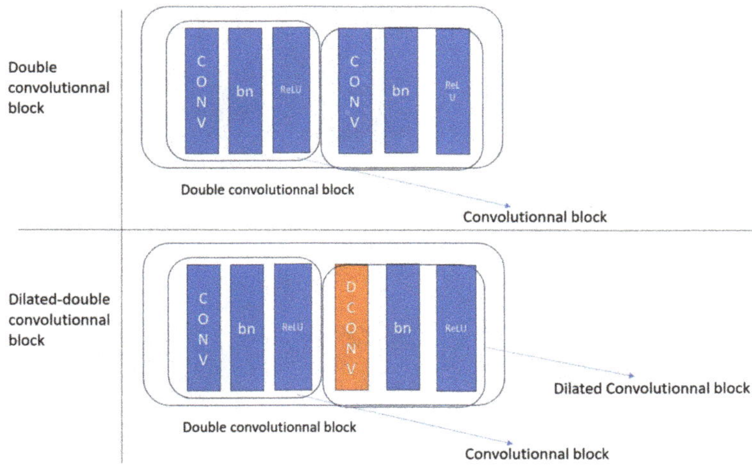

Fig. 1.5 The double convolutional block consists of two blocks with a convolutional layer followed by a batch normalization (BN) and a ReLU activation; the second block uses dilated convolutions

depth of the last layer in the decoding path. Subsequently, a sigmoid function is applied to map the output values to a range between 0 and 1. The output of this last layer consists of two images with the same resolution as the input image. Using a threshold, the pixels of the input image are divided into two classes, 0 and 1, to produce the final segmentation.

The U-Net base model allows an end-to-end training of a deep neural network for classification with a relatively small database thanks to the skip connections and to the moderate complexity of the model. The hyperparameters, called *filters*, represent the number of convolutional filters in each convolutional layer of the first double convolutional block. The depth represents the convolutional layer number fixed similarly to the original U-Net model. Then the number of filters is doubled at each double convolutional block, similar to the U-Net model. The number of filters is grown at each block of the network because the features extracted are more complex at each layer and the combination of these features is more interesting at deeper layers than in the early layers of the network. Both the VGG-16 models and U-Net models use 64 filters in the first layer. After conducting various simulations, we consistently observed that using 64 filters degraded the performance of the models and the segmentation and classification of GSS. Therefore, we use 32 filters considering the computational cost, especially the memory size of the deep neural network.

The best segmentation model of salivary glands provided a mean dice of 0.82 with a standard deviation of 0.21 on the test database. Figure 1.6 presents samples from the test set with the expert annotation and the predicted segmentation. This shows that the model had not over-fitted the data and learned real features. This can

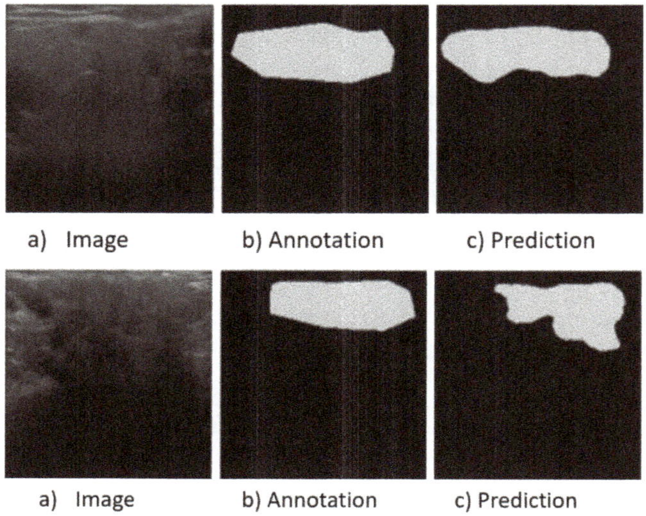

Fig. 1.6 Examples of segmentation. (**a**) Good predicted segmentation. (**b**) Predicted segmentation with a different shape but a dice>0.9

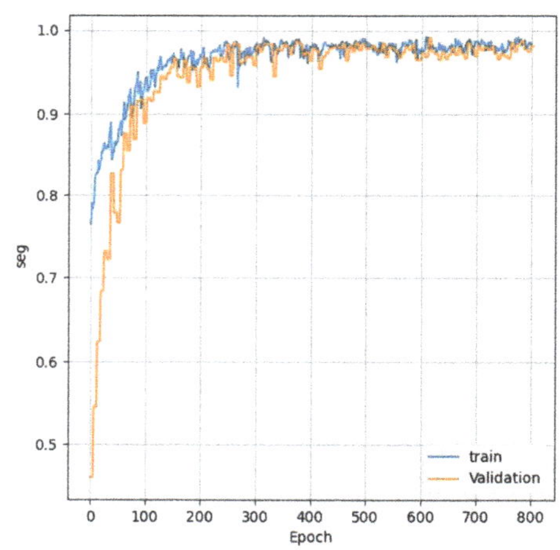

Fig. 1.7 Evolution of the dice values on the training set at each epoch of the optimization and on validation set every five epochs

be observed on the graph that represents the evolution of the mean dice obtained on the training and validation set during the training (see Fig. 1.7).

To conclude on this segmentation task, we obtain a satisfying mean dice value for the segmentation of the salivary glands in ultrasound imaging, which provides a region of interest for the computation of handcrafted features for the classification of the GSS.

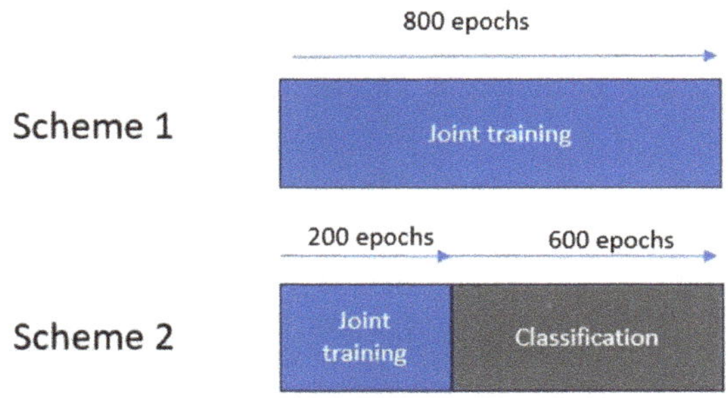

Fig. 1.8 Training schemes: scheme 1 represents a joint training performed during 800 epochs, whereas scheme 2 represents a joint training followed by a specialization in classification

1.6.3 Multiphase Joint Training Scheme

Hereinafter, we describe the multiphase joint-training scheme used to train the deep neural network. The joint classification and segmentation (or multitask model) model has already been introduced in [22]. However, our network model is inspired by the U-Net model. We built a joint segmentation and classification network, and we proposed a novel training scheme which consists either (see Fig. 1.8):

- In a single-phase training entirely based on the multitask loss function
- In several phases mixing joint training and specific training on the classification or segmentation

The training duration is fixed to 800 epochs.[8] Figure 1.8 represents a one-phase training applied in the multitask setting. The second scheme displayed is the two-phase joint training, with 200 epochs of multitask training and 600 epochs of a specific classification training phase. The final model used for the classification has the same number of parameters as a network trained on classification only. We used this regularization to push the network to produce high-level task-relevant features using low-level features built in the first phase, thus fine-tuning the network more specifically.

1.6.4 Ydnet Architecture

The joint classification and segmentation model is built from the base segmentation model presented in Sect. 1.6.2. We call the classification branch the linear neural layers added to the basic segmentation model. Figure 1.9 displays the base segmentation model with the classification branch. In this figure, we draw links between the layers of the segmentation model and the input of the classification

[8] An epoch is performed when the model has been optimized with every sample of the training dataset. Each iteration of the gradient descent in the epoch is performed on a subset of the dataset in order to fit into the memory available in the processing unit.

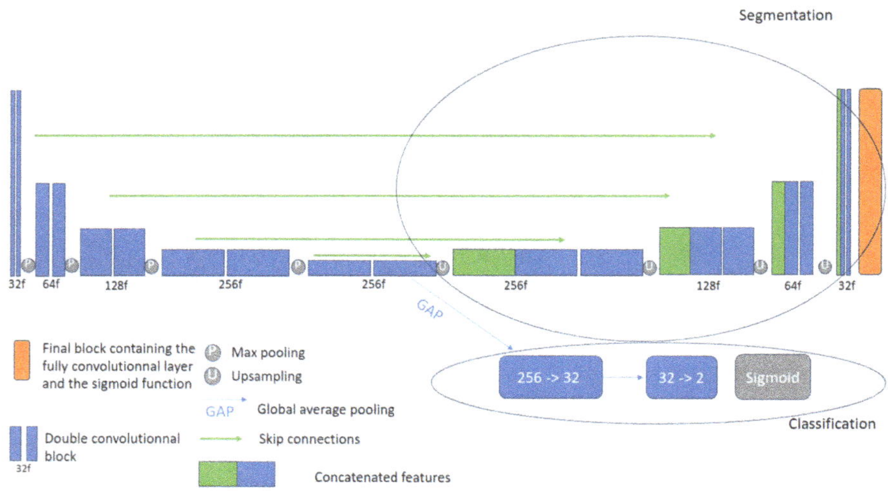

Fig. 1.9 Our joint classification and segmentation model is based on the U-Net model with an additional classification path using a global average pooling and linear layers

branch. This layer, which we call the bottleneck layer, is situated at the end of the encoding path of the segmentation model and before the decoding path.

The classification part incorporates a global average pooling operation. This operation takes the output of the bottleneck layer as input and produces a vector with a size equivalent to the number of filters at the bottleneck layer. Then two linear layers are applied, followed by the sigmoid function that produces the final class probabilities. The probabilities are then thresholded above and below 0.5 to produce the final binary classification. We call the model Ydnet, where "Y" stands for the two output branches and "d" for the dilated convolutions. The encoding part of the model uses the same blocks as the segmentation model.

1.6.5 Loss Function and Hyperparameters

The segmentation loss function is based on a sum of cross-entropy and dice score. Let x_i and y_i be the binary predicted label and ground truth label for a pixel $i \in [1, n]$, respectively, where n stands for the number of pixels. The segmentation cross-entropy loss is defined as:

$$l_{seg_{ce}} = -\sum_{i=1}^{n} (y_i log(x_i) + (1 - y_i) log(1 - x_i)) \tag{1.3}$$

Let x_{il} and y_{il} be, respectively, the binary predicted label and ground truth label for a voxel $i \in [1, m]$, where m stands for the number of voxels, and let l be the class label with $l \in \{0, 1\}$. The binary dice is defined as:

$$l_{dice} = 2\frac{\sum_{l=0}^{1}\sum_{i=1}^{m} x_{il} y_{il}}{\sum_{l=0}^{1}\sum_{i=1}^{m}(x_{il} + y_{il})} \tag{1.4}$$

Let x and y be the class prediction and the label of a sample, respectively; the classification categorical cross-entropy is computed as follows:

$$l_{ce} = -\sum_{i=1}^{n}(y\log(x) + (1-y)\log(1-x)) \tag{1.5}$$

We should emphasize that the classification cross-entropy loss is computed for every sample, while the segmentation cross-entropy is calculated considering each pixel within every sample. During the first phase, the cross-entropy is used as classification loss and a weighted sum of a segmentation cross-entropy loss, and the dice loss produces the segmentation loss. Then, during the second phase of 600 epochs, we only use the classification cross-entropy to train the network. The total loss function is evaluated as a weighted sum where the weight of the classification cross-entropy is denoted as w_{ce}, the weight of the segmentation cross-entropy loss as $w_{seg_{ce}}$, and the term of the dice loss as $w_{seg_{dice}}$.

$$L_{total} = w_{ce}L_{classif} + w_{seg_{dice}}L_{segdice} + w_{seg_{ce}}L_{segCE} \tag{1.6}$$

Concerning the dice coefficient, we selected various values. These values impact the magnitude of gradients during the training and are selected to prevent any loss gradient from dominating the others. To improve the network regularization, we added a weighted semi-supervised loss function that forces the model to produce low-entropy predictions, thus increasing the confidence in the predictions of the classifier:

$$L_{total} = w_{ce}L_{classif} + w_{seg_{dice}}L_{segdice} + w_{seg_{ce}}L_{segCE} + w_{ssl}L_E \tag{1.7}$$

1.6.6 Simulations

The *batch size* corresponds to the number of samples used for each optimization step. The batch size and the number of *filters* should be selected according to the GPU memory. We should highlight that both the batch size and the learning rate impact the optimization process. A larger batch size can provide more accurate gradient estimates. The batch size was set at 4, the maximal possible value considering the memory issues. Additionally, we observed that a larger number of filters lead to more risk of over-fitting. The number of filters in the original U-Net is set to 64. However, in our project on classification and segmentation tasks, we found that 32 filters produced better results. To learn the weights of the network, the optimization algorithm was based on the Adam optimizer, which adds momentum to the classical stochastic gradient descent algorithm [18]. The learning rate, which weights the optimization steps, was set to 10^{-5} which did not show optimization issues. The training of the deep learning models was performed using the framework PyTorch [24] with a graphics processing unit (GPU) Nvidia volta V100 with a memory of 32 Go. Table 1.4 shows a base set of hyperparameters.

Hereinafter, we present and compare the two-phase training results to the ones of classical training with classification loss. Based on our database, we compare our results to the ones of a machine learning classifier. We also consider the results of a model trained on our database and used to detect GSS on Harmonics. The best accuracy was obtained with eight augmentations at each batch; see Fig. 1.10. Additionally, there is a large gap between the accuracy without and with augmentations. We used the following augmentation methods: sharpen, piecewise affine, linear contrast, Gaussian blur, additive Gaussian noise, edge detection, dropout, and elastic deformation.

We performed a twofold cross-validation with two launches for each set of hyperparameters to test the various training schemes. Table 1.5 shows that the two-phase and the one-phase joint trainings on the classification and segmentation

Table 1.4 List of hyperparameters tested for the comparison of the multitask multiphases model with classical training

Hyperparameters	Values
Filters at first layer	32, 64
Augmentations	0, 1, 4, 8
Training scheme	[2 phases, class]
Loss coefficients	{[10, 0.2, 0.1], [10, 0.2, 0], [1, 1, 0], [10, 0.1, 0], [1, 0.5, 0], [10, 0.01, 0], [10, 0.3, 0]}
Normalization	["no," "standard"]
Image shape (width, height)	{(128, 128), (192, 192)}

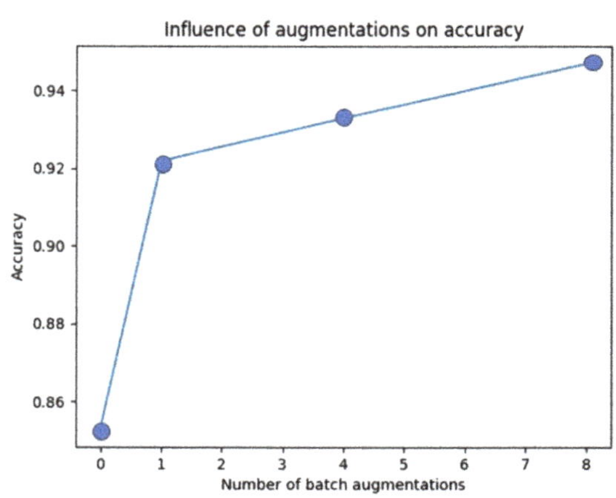

Fig. 1.10 Influence of the number of augmentations on the accuracy with measures only for 0, 1, 4, or 8 augmentations

Table 1.5 Twofold accuracy for two phases or classification phase only without normalization averaged over various image shapes

Training phases	Norm	Accuracy
1-phase joint	no	0.91
2-phase joint	no	0.91
classification	no	0.89

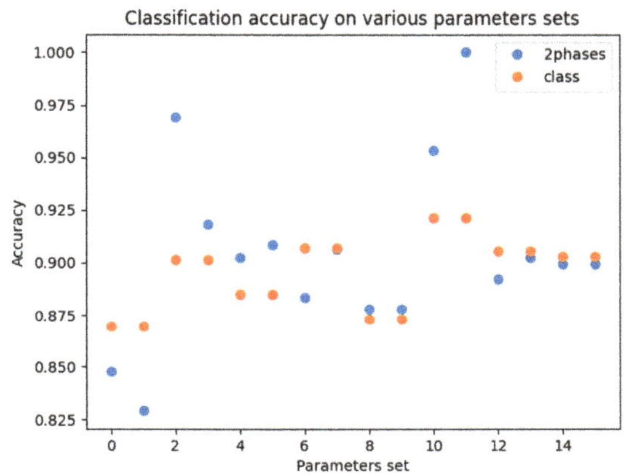

Fig. 1.11 Classification accuracy for all sets of hyperparameters

produce similar results with a mean accuracy 0.02 higher (or 2% higher) than the results of only the classification task. However, when comparing the results using a base of 32 filters at the first layer against 64 filters, we observe differences between the two methods. It seems that the regularization provided by the joint classification and segmentation improves the metrics.

The results, obtained with various combinations of loss function weights $w_{\text{seg\_dice}}$ and $w_{ce} = 1$, showed similar results. We observe a correlation between a slight decrease in the accuracy and a decrease in the segmentation weight coefficient. The best results are obtained with neighboring values for the classification of cross-entropy and the segmentation dice. However, lower values are obtained when adding the segmentation cross-entropy loss or when using a small coefficient for the segmentation dice loss. Over all hyperparameter sets, accuracy differs slightly between the two methods; see Fig. 1.11. The three best results and the two lowest precisions appeared with the two-phase model, but it is hard to draw a conclusion from such a slight variation in accuracy.

The maximum accuracy on one fold of the cross-validation is obtained with the two-phase model, without normalization, but with a classification loss coefficient of 1 and a dice segmentation loss coefficient of 0.5. The best classification model based on handcrafted features provided an accuracy of 0.84 with a sensitivity of 0.70 and a specificity of 0.90. We should notice that the sensitivity provided by this model may be relatively low for clinical applications, whereas the deep learning models provided almost perfect results. To conclude, the best deep learning model, obtained with a deep neural network trained in two phases, provided an accuracy 0.16 higher that the best model based on a logistic regression and 23 selected features.

1.7 Application on an External Database: HarmonicSS

We performed two types of experiments on the HarmonicSS database. First, we performed predictions on HarmonicSS with a model trained on our database with various scores and cross-validation. Then, we fine-tuned the same model on separated three subsets of HarmonicSS used for training, testing, and validation. We used the same metrics as previously described to measure the performance. By applying the model trained over GSID to HarmonicSS, the best accuracy obtained was 0.83 using the devita1 score and an image size of (192, 192); see Table 1.6.

The results obtained on original HarmonicSS images without preprocessing were very low, with an accuracy of 0.55 obtained with devita0. A preprocessing method called "Adapted," which reshaped images adapted to all different input shapes, performed a lower accuracy than reshaping all images to (192,192). The accuracy decreased from 0.92 on GSID to 0.83 on HarmonicSS. The direct predictions on HarmonicSS compared with devita0, using the trained model, produced an accuracy of 0.76, with a low sensibility and a high specificity. The fine-tuning of the same model performed using devita0 classification produced an accuracy of 0.87, which is higher than the one obtained with direct predictions with a much higher sensibility of 0.90 against 0.67. The accuracy on the devita1 score is higher than one on the devita0 score on the direct prediction (Table 1.7).

Table 1.6 GSS detection with two-phase model trained on GSID and predicted on HarmonicSS with a cross-validation

Score	Image shape	Accuracy	Sensitivity	Specificity
devita0	(192,192)	0.75	0.59	0.92
devita0	Original	0.55	0.58	0.51
devita0	Adapted	0.75	0.57	0.94
devita0	(192,144)	0.76	0.67	0.96
devita1	(192,192)	0.83	0.72	0.90

Table 1.7 Fine-tuning on devita0 and devita1 on HarmonicSS compared to direct prediction

Score	Model	Accuracy	Sensitivity	Specificity
devita0	Direct	0.76	0.67	0.96
devita1	Direct	0.83	0.72	0.90
Eular	Direct	0.94	0.95	0.77
Omeract	Direct	0.74	0.72	0.78
devita0	Fine-tuned	0.87	0.90	0.85
devita1	Fine-tuned	0.93	0.92	0.93

1.8 Conclusion

Our project focuses on the analysis of various Gougerot-Sjögren syndrome (GSS) detection methods using ultrasound imaging. Our study considers the optimal classifier, normalization techniques, bin-width settings, and the most pertinent radiomic features. Additionally, we conducted a comparative evaluation of two distinct deep neural network training strategies. This comprehensive assessment encompassed hyperparameter optimization and the exploration of training loss coefficients. These combined efforts enabled us to draw insightful comparisons between deep learning and traditional machine learning approaches while also evaluating the robustness of the model on an unseen database.

Our segmentation model provided a dice of 0.82 with a standard deviation of 0.21. In [33], the dice of the inter-observer agreement reached 0.86 and the dice of the intra-observer agreement reached 0.91 on a dataset containing 1184 SGUS. Thus, the value of 0.82 obtained with a smaller database is sufficient to extract a region of interest for the extraction of handcrafted features. The segmentation model trained on GSID was used to train models based on handcrafted features on the HarmonicSS database without the ground truth mask. The best accuracy, 0.72, was obtained with the regression model on an Std norm of features and without image normalization. The prediction of the best model trained on GSID (i.e., SVM_Std) and used on HarmonicSS provided an accuracy of 0.63, a sensibility of 0.50, and a specificity of 0.77. These results are low compared to ones obtained with an adaptation of the deep learning model to HarmonicSS.

The classification tests using radiomic features showed that the feature selection,based on the random forest classifier, improved the accuracy. This highlights the importance of feature selection in the training phase of machine learning classifiers on radiomic features. The feature selection method has a large impact on accuracy. The best parameters for the two-phase training were obtained with an input image shape of (128, 128) and without any image normalization. A dice loss coefficient close to the classification loss coefficient shows an accuracy of 1. Based on accuracy, the best results are obtained in the salivary glands database, with the two-phase model with a twofold cross-validated accuracy of 1.0 compared to the classification based on the radiomic features with 0.84. These results show that both methods are relevant for the detection of GSS, but the deep learning model provides 16% higher performances. We should highlight that a deep learning model is required to produce the segmentation used in the handcrafted feature extraction. The latter procedure increases the complexity, computation efforts, and required time compared to the deep learning–based method. The deep network trained on GSID was found to be well adapted to an unseen environment on the open-source HarmonicSS database with an accuracy of 0.83 based on the devita1 score. Furthermore, better results were obtained using HarmonicSS with devita0 and devita1 scores using a fine-tuning model trained on GSID.

Our experiments showed that the deep learning models gain in accuracy due to the larger possibility of encoding shifts in images and environment. Indeed, the deep learning model provided a strong classification for the Gougerot-Sjögren syndrome. A single model on a single dataset split provided an accuracy of 1.0 on the test set. We also noticed that the maximal classification accuracy score was higher with the two-phase model, which denotes that this training method can lead to better models but is less consistent.

Acknowledgments This research project is supported by the French Clinical Research Infrastructure Network on Venous Thrombo-Embolism (FCRIN INNOVTE). The authors would also like to acknowledge Brest University Hospital and ENSTA Bretagne for their support.

References

1. Berthomier, T., Mansour, A., Bressollette, L., Le Roy, F., Mottier, D.: Venous blood clot structure characterization using scattering operator. In: International Conference on Frontiers of Signal Processing (ICFSP), pp. 73–80 (2016)
2. Berthomier, T., Mansour, A., Bressollette, L., Le Roy, F. Dominique Mottier, Fréchier, L., Hermenault, B.: Scattering operator and spectral clustering for ultrasound images: application on deep venous thrombi. Int. J. Biomed. Biol. Eng. **11**(11), 630–637 (2017)
3. Brandt, J.E., Priori, R., Valesini, G., Fairweather, D.L.: Sex differences in sjögren's syndrome: a comprehensive review of immune mechanisms. Biol. Sex Differ. **6**, 1–13 (2015)
4. Cornec, D., Jousse-Joulin, S., Costa, S., Marhadour, T., Marcorelles, P., Berthelot, J.M., Hachulla, E., Hatron, P.Y., Goeb, V., Vittecoq, O., et al.: High-grade salivary-gland involvement, assessed by histology or ultrasonography, is associated with a poor response to a single rituximab course in primary sjögren's syndrome: data from the tears randomized trial. PLoS One **11**(9), e0162787 (2016)
5. De Vita, S., Lorenzon, G., Rossi, G., Sabella, M., Fossaluzza, V.: Salivary gland echography in primary and secondary sjögren's syndrome. Clin. Exp. Rheumatol. **10**(4), 351–356 (1992)
6. Deng, J., Dong, W., Socher, R., Li, L. J., Li, K., Fei-Fei, L.: Imagenet: a large-scale hierarchical image database. In: 2009 IEEE Conference on Computer Vision and Pattern Recognition, pp. 248–255. IEEE, New York (2009)
7. Ghannam, M.G., Singh, P.: Anatomy, Head and Neck, Salivary Glands. StatPearls Publishing, Treasure Island (FL) (2022)
8. Glorot, X., Bengio, Y.: Understanding the difficulty of training deep feedforward neural networks. In: Proceedings of the Thirteenth International Conference on Artificial Intelligence and Statistics, pp. 249–256. JMLR Workshop and Conference Proceedings (2010)
9. Goules, A.V., Exarchos, T.P., Pezoulas, V.C., Kourou, K.D., Venetsanopoulou, A.I., De Vita, S., Fotiadis, D.I. and Tzioufas, A.G.: Sjögren's syndrome towards precision medicine: the challenge of harmonisation and integration of cohorts. Clin. Exp. Rheumatol **37**(Suppl 118), S175–84 (2019)
10. Haralick, R.M., Shanmugam, K., Dinstein, I.H.: Textural features for image classification. IEEE Trans. Syst. Man Cybern. **SMC-3**(6), 610–621 (1973)
11. Hayter, S.M., Cook, M.C.: Updated assessment of the prevalence, spectrum and case definition of autoimmune disease. Autoimmun. Rev. **11**(10), 754–765 (2012)
12. He, K., Zhang, X., Ren, S., Sun, J.: Deep residual learning for image recognition. In: Proceedings of the IEEE Conference on Computer Vision and Pattern Recognition, pp. 770–778 (2016)
13. Hochreiter, S., Schmidhuber, J.: Simplifying neural nets by discovering flat minima. Adv. Neural Inf. Proces. Syst. **7**, 529–536 (1994)

14. James-Goulbourne, T., Murugesan, V., Kissin, E.Y.: Sonographic features of salivary glands in Sjögren's syndrome and its mimics. Curr. Rheumatol. Rep. **22**, 1–9 (2020)
15. Jousse-Joulin, S., d'Agostino, M.A., Nicolas, C., Naredo, E., Ohrndorf, S., Backhaus, M., Tamborrini, G., Chary-Valckenaere, I., Terslev, L., Iagnocco, A., et al.: Video clip assessment of a salivary gland ultrasound scoring system in Sjögren's syndrome using consensual definitions: an omeract ultrasound working group reliability exercise. Ann. Rheum. Dis. **78**(7), 967–973 (2019)
16. Jung, A.B., Wada, K., Crall, J., Tanaka, S., Graving, J., Reinders, C., Yadav, S., Banerjee, J., Vecsei, G., Kraft, A., Rui, Z., Borovec, J., Vallentin, C., Zhydenko, S., Pfeiffer, K., Cook, B., Fernández, I., De Rainville, F.-M., Weng, C.-H., Ayala-Acevedo, A., Meudec, R., Laporte, M., et al.: imgaug. https://github.com/aleju/imgaug (2020). Online; Accessed 01-Feb-2020
17. Kassan, S.S., Moutsopoulos, H.M.: Clinical manifestations and early diagnosis of Sjögren syndrome. Arch. Intern. Med. **164**(12), 1275–1284 (2004)
18. Kingma, D.P., Ba, J.: Adam: A method for stochastic optimization. In: Bengio, Y., LeCun, Y., (eds.). 3rd International Conference on Learning Representations, ICLR 2015, San Diego, CA, USA, May 7–9, 2015, Conference Track Proceedings (2015)
19. Kise, Y., Shimizu, M., Ikeda, H., Fujii, T., Kuwada, C., Nishiyama, M., Funakoshi, T., Ariji, Y., Fujita, H., Katsumata, A. and Yoshiura, K., Ariji, E.: Usefulness of a deep learning system for diagnosing Sjögren's syndrome using ultrasonography images. Dentomaxillofacial Radiol. **49**(3), 20190348 (2020). PMID: 31804146
20. Kumar, V., Gu, Y., Basu, S., Berglund, A., Eschrich, S. A., Schabath, M. B., Gillies, R.J. Forster, K., Aerts, H.J.W.L., Dekker, A., Fenstermacher, D., et al.: Radiomics: the process and the challenges. Magn. Reson. Imaging **30**(9), 1234–1248 (2012)
21. Le Mélédo, G., Jousse-Joulin, S.: Échographie des glandes salivaires en rhumatologie. Revue du Rhumatisme Monographies **88**(4), 274–278 (2021). Crâne et face
22. Mehta, S., Mercan, E., Bartlett, J., Weaver, D., Elmore, J.G., Shapiro, L.: Y-net: joint segmentation and classification for diagnosis of breast biopsy images. In: International Conference on Medical Image Computing and Computer-Assisted Intervention, pp. 893–901. Springer, Berlin (2018)
23. Nair, V., Hinton, G.E.: Rectified linear units improve restricted boltzmann machines. In: Proceedings of the 27th International Conference on International Conference on Machine Learning (ICML'10), pp. 807–814. Omnipress, Madison, WI, USA (2010)
24. Paszke, A., Gross, S., Massa, F., Lerer, A., Bradbury, J., Chanan, G., Killeen, T., Lin, Z., Gimelshein, N., Antiga, L., Desmaison, A., Kopf, A., Yang, E., DeVito, Z., Raison, M., Tejani, A., Chilamkurthy, S., Steiner, B., Fang, L., Bai, J., Chintala, S.: Pytorch: An imperative style, high-performance deep learning library. In: Wallach, H., Larochelle, H., Beygelzimer, A., d'Alché-Buc, F., Fox, E., Garnett, R. (eds.) Advances in Neural Information Processing Systems, vol. 32, pp. 8024–8035. Curran Associates, Inc., New York (2019)
25. Price, E.J., Tappuni, A.R., Sutcliffe, N.: Mimics of Sjögren's syndrome. In: Oxford Textbook of Sjögren's Syndrome. Oxford University, Oxford (2021)
26. Ronneberger, O., Fischer, P., Brox, T.: U-net: Convolutional networks for biomedical image segmentation. In: International Conference on Medical Image Computing and Computer-Assisted Intervention, pp. 234–241. Springer, Berlin (2015)
27. Rumelhart, D.E., Hinton, G.E., Williams, R.J., et al.: Learning Internal Representations By Error Propagation (1985)
28. Shiboski, C.H., Shiboski, S.C., Seror, R., Criswell, L.A., Labetoulle, M., Lietman, T.M., Rasmussen, A., Scofield, H., Vitali, C., Bowman, S.J., et al.: 2016 american college of rheumatology/european league against rheumatism classification criteria for primary sjögren's syndrome: a consensus and data-driven methodology involving three international patient cohorts. Ann. Rheum. Dis. **76**(1), 9–16 (2017)
29. Soille, P.: Erosion and dilation. In: Morphological Image Analysis, pp. 63–103. Springer, Berlin (2004)
30. Sun, C., Wee, W.G.: Neighboring gray-level dependence matrix for texture classification. Comput. Vision, Graphics Image Processing **23**(3), 341–352 (1983)

31. Thibault, G., Angulo, J., Meyer, F.: Advanced statistical matrices for texture characterization: application to cell classification. IEEE Trans. Biomed. Eng. **61**(3), 630–637 (2013)
32. Vukicevic, A.M., Milic, V., Zabotti, A., Hocevar, A., De Lucia, O., Filippou, G., Frangi, A.F., Tzioufas, A., De Vita, S., Filipovic, N.: Radiomics-based assessment of primary sjögren's syndrome from salivary gland ultrasonography images. IEEE J. Biomed. Health Inform. **24**(3), 835–843 (2019)
33. Vukicevic, A.M., Radovic, M., Zabotti, A., Milic, V., Hocevar, A., Callegher, S.Z., De Lucia, O., De Vita, S., Filipovic, N.: Deep learning segmentation of primary Sjögren's syndrome affected salivary glands from ultrasonography images. Comput. Biol. Med. **129**, 104154 (2021)
34. Weszka, J.S., Dyer, C.R., Rosenfeld, A.: A comparative study of texture measures for terrain classification. IEEE Trans. Syst. Man Cybern. **SMC-6**(4), 269–285 (1976)
35. Zabotti, A., Zandonella Callegher, S., Tullio, A., Vukicevic, A., Hocevar, A., Milic, V., Cafaro, G., Carotti, M., Delli, K., De Lucia, O., et al.: Salivary gland ultrasonography in Sjögren's syndrome: a European multicenter reliability exercise for the harmonicss project. Front. Med. **7**, 581248 (2020)
36. Zwanenburg, A., Vallières, M., Abdalah, M.A., Aerts, H.J.W.L., Andrearczyk, V., Apte, A., Ashrafinia, S., Bakas, S., Beukinga, R.J., Boellaard, R., et al.: The image biomarker standardization initiative: standardized quantitative radiomics for high-throughput image-based phenotyping. Radiology **295**(2), 328–338 (2020)

Chapter 2
Deep Learning Classification of Venous Thromboembolism Based on Ultrasound Imaging

A. Olivier, A. Mansour, C. Hoffmann, L. Bressollette, and B. Clement

Abstract Venous thromboembolism (VTE) occurs when a blood clot forms in a vein. According to the US National Institutes of Health, VTE affects 0.13% of men and around 0.11% of women in the United States every year, i.e., about 400 000 people per year. VTE includes deep vein thrombosis (DVT) and pulmonary embolism (PE). DVT is linked to the obstruction of a deep vein by a blood clot, usually in the lower leg, thigh, or pelvis. Whereas pulmonary embolism (PE) results from the migration of the blood clot toward a pulmonary artery. The objective of our project is to evaluate the possibility of predicting a PE based on ultrasound (US) images. It should be emphasized that there is no medical expertise for the detection of PE from these images. We proposed two methods: the first is based on the extraction of texture descriptors and the second relies on deep learning models. We developed a learning scheme for deep neural networks based on a joint training on a classification and segmentation task, and then a specialization of the network on the classification task. Alternatively, we built a model combining images and clinical data. Beyond the techniques used, significant work has been carried out to sort the database studied and select images. We obtained conclusive accuracy on the detection of PE.

A. Olivier · A. Mansour (✉)
ENSTA, Lab-STICC UMR 6285 CNRS, Brest, France
e-mail: aurelien.olivier@ensta-bretagne.org; mansour@ieee.org

C. Hoffmann · L. Bressollette
GETBO UMR 13-04 CHRU Cavale Blanche, Brest, France
e-mail: luc.bressollette@wanadoo.fr

B. Clement
ENSTA, Lab-STICC UMR 6285 CNRS, Brest, France

CROSSING IRL CNRS, Adelaide, SA, Australia
e-mail: benoit.clement@ensta-bretagne.fr

F. Dornaika et al. (eds.), *Advances in Data Clustering*,
https://doi.org/10.1007/978-981-97-7679-5_2

Keywords Machine learning · Deep learning · Texture analysis · Radiomics · Classification · Multi-supervision · Ultrasound imaging · Gougerot-Sjögren Syndrome

2.1 Introduction

Venous thromboembolism (VTE) is a serious and common ailment, defined by the growth of a clot in a vein and may be associated with a potential migration of this thrombus to a pulmonary artery. VTE represents the third cause of death after the cardiovascular disease and cancer [7] and encompasses superficial thrombosis, muscular thrombosis, deep venous thrombosis (DVT) and pulmonary embolism (PE), which can be isolated or associated. DVT develops within a deep vein of the body, which can detach and provoke a PE by obstructing the flow of blood in a pulmonary artery. In a study examining the causes of death within a general hospital patient population, PE was identified as the cause of death in 10% of cases [20]. A major challenge in the diagnosis and treatment of a DVT lies in the diversity of phenotypes of patients affected by VTE which can be provoked (by known factors as an immobilization surgery or trauma) or unprovoked which may indicate that the patient has an increased tendency to develop a clot. Thus, the etiology of VTE plays an important role in the analysis of recidivism risks of PE.

The interest in the detection of PE from ultrasound (US) images of a DVT is twofold. First, an accurate detection would show that there is information in the texture or the shape of the clot, relative to the risk of PE, and it would be a step to a gradual classification of the risk of PE. Second, it would be interesting for the practitioners to be able to detect a PE using only US of DVT. While the detection of PE is already practiced clinically with other imaging methods, US is cheaper, faster, more easily available, and noninvasive. Additionally, in the most severe cases, a more aggressive treatment can be recommended following the examination, with a hospitalization or a thrombolysis. We should highlight that the images that we use in this project are selected by the clinicians when performing the compression ultrasound study in order to diagnose the presence of a DVT. This implies that the data available is a snippet of the original US examination and provides less information, which makes the task more complex. For example, it is sometimes impossible for a practitioner to locate the blood clot with a single image. Besides, US suffers from noise and artifacts.

In order to perform segmentation and classification tasks, we used two different methods to extract features using deep learning models and handcrafted (HC) features. HC features are identified and computed features, while deep learning features are automatically obtained during the learning process. The segmentation can be produced either by an expert or by a deep learning model. The segmentation consists in the classification of every pixel of an image instead of the classification of the whole image in one or several classes. Long et al. [13] proposed a model called fully convolutional network (FCN) with a fully convolutional layer and a learned "deconvolutional" path, in order to up-sample the feature maps at the original input size and thus perform a fine pixel-wise prediction. The U-Net [19] model reached better performance on the International Symposium on Biomedical Imaging (ISBI) cell tracking challenge dataset. U-Net is based on three novelties compared to FCN: the skip connections, the double convolutional blocks and a deconvolutional path. The skip connections copy the feature maps from the output of each double convolutional block and concatenate it with the up-sampled feature maps of the corresponding resolution in the deconvolutional path. This concatenation enhances the capacity of the model to "localize." The double convolutional block consists in stacking two convolutional layers with normalization and activations at each step of the down-sampling path. The deconvolutional path consists of a learned up-sampling path with 3×3 convolutions in order to output feature maps with the same shape as the input.

While many improvements were proposed on the basic U-Net model, either with an injection of attention modules in the model [15], Swin transformer blocks [2], or using a ResNet model as a backbone [26]; the original U-Net model is still competitive in biomedical image segmentation and used as a baseline to compare new developed methods [9]. In our experiments, we use a model derived from U-Net and perform several variations on the model in our experiments.

In the following sections and due to a page length limit, we briefly explain our approach in Sect. 2.2. Section 2.3 provides insights into different experiments conducted. Besides, it presents and discusses some simulation results. We end our chapter with a general conclusion.

2.2 Our Proposed Approach

It is well known that the database is an important aspect of any deep learning approach. In our project, we used a database called **EDITH** and collected during 10 years at French University hospitals. EDITH provides ultrasound images with clinical metadata.

2.2.1 Concatenation of Image and Clinical Data

According to medical experts, we selected five potential clinical factors to be included in the clinical and image fusion model: sex, age, presence of cancer, body mass index (BMI) and hemoglobin (HB). Later on, we developed two methods to make better use of this available supplementary data:

- First, we perform a concatenation of the clinical data vector, with a global average pooling of the last layer of the encoding path of the deep learning model; see

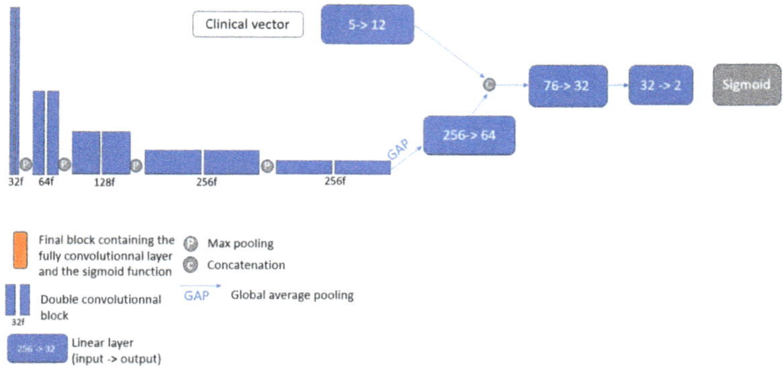

Fig. 2.1 The fusion model with the concatenation of image features and clinical features

Fig. 2.1. Finally, a cross-entropy loss is used to train the model as shown in Fig. 2.1, with an Adam optimizer as a basis [11].

- Inspired by [16], the second method consists in placing an attention block[1] within the network, in order to scale and shift feature maps according to the clinical data values. Additionally, we introduce a transfer learning method to retrain the 16-layer vision transformer model (VIT16) [6]. Given a feature map x of size (n_z, n_x) and the clinical data vector z of size n_z, this module called metablock can be written as follows, considering σ as the sigmoid function:[2]

$$x_n = \sigma(\tanh(f(z) \cdot x) + g(z)) \tag{2.1}$$

The functions f and g are built with weights matrix W_f and W_g and bias weights w_{0_f} and w_{0_g} of size (n_z, n_x). The functions can be written as follows:

$$f(z) = W_f^T z + w_{0_f} \tag{2.2}$$

$$g(z) = W_g^T z + w_{0_g} \tag{2.3}$$

Figure 2.2 gives a visual example of the selection of the features applied with a vector containing clinical information. This method is applied to the features of the last convolutional layer of the model, just before the linear layers.

Very deep neural networks, such as VIT16 [6] which contains 86 billion parameters, are initially pretrained on extensive datasets like ImageNet. The weights of these pretrained models are then made available on major frameworks used for

[1] CNN models first increased the performance on benchmark datasets, with ResNet providing strong performances on benchmarks such as ImageNet. Many works then focused on trying to convert a new kind of models called transformers that provided large improvements on Natural language processing tasks. Those models incorporate self-attention modules [4, 17, 23].

[2] Sigmoid function: $\sigma(x) = \frac{1}{1+e^{-x}}$.

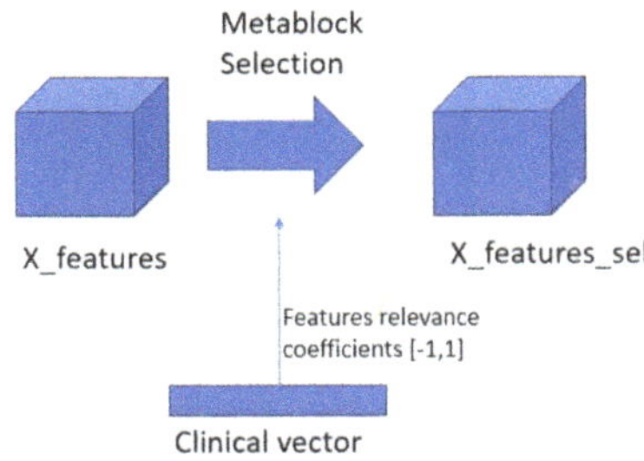

Fig. 2.2 Selection of features with the metablock from a clinical data vector

training deep neural networks, such as PyTorch[3] [18]. However, the majority of models are trained on red, blue and green (RGB) images, and as of now, there are no models pretrained on large-scale grayscale databases. Hence, when employing transfer learning from pretrained models, it is required to adapt these pretrained RGB models for grayscale tasks. We fine-tuned the VIT-16 base model with pretrained weights obtained through a two-step process. Initially, weakly supervised pretraining was conducted on a dataset comprising 3.6 billion publicly available images, collected with corresponding hashtags (27k hashtags). Subsequently, the model was fine-tuned on the ImageNet dataset.

2.2.2 Architecture Details

As previously mentioned, our approach can be divided into two major tasks: segmentation and classification. The global architecture can be seen in Fig. 2.3. That figure displays the base segmentation model with the classification branch. It draws the link between one of the layers of the segmentation model and the input of the classification branch. This layer, called the bottleneck layer, is situated at the end of the encoding path of the segmentation model and before the decoding path. The classification part incorporates a global average pooling operation. This operation takes the output of the bottleneck layer as input and produces a vector with a size equivalent to the number of filters at the bottleneck layer. Then, two linear layers are applied, followed by the sigmoid function that produces the final class probabilities. These probabilities are then threshold above and below 0.5 to produce the final binary classification. We call the model Ydnet, where "Y" stands for the two output branches and "d" for dilated convolutions. The encoding part of the model uses the same blocks as the segmentation model.

[3] The pretrained weights can be found in the following URL https://pytorch.org/.

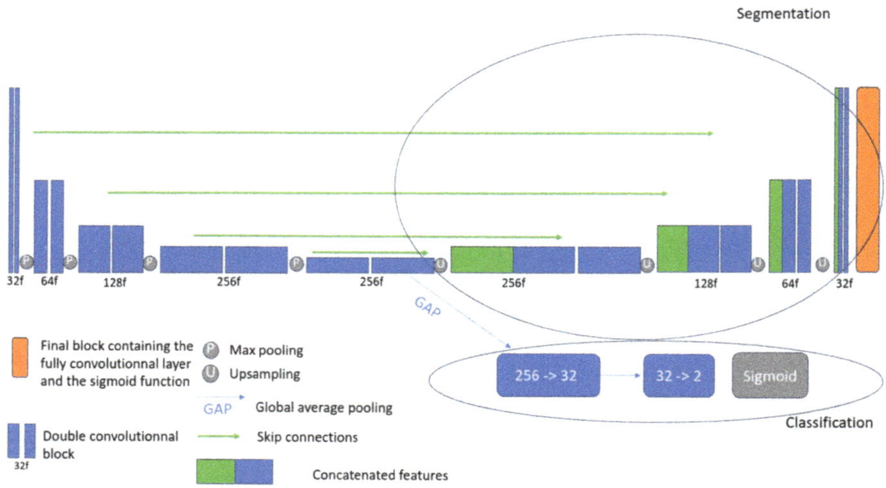

Fig. 2.3 The figure represents the proposed model for the joint classification and segmentation. The model is based on the U-net model with an additional classification path using a global average pooling and linear layers

2.2.3 Loss Function and Hyperparameters

The segmentation loss function is based on a sum of cross-entropy and dice score. Let x_i and y_i be, respectively, the binary predicted label and ground truth label for a pixel $i \in [1, n]$, where n stands for the number of pixels. The segmentation cross-entropy loss is defined as:

$$L_{ce} = -\sum_{i=1}^{n} (y_i log(x_i) + (1 - y_i) log(1 - x_i)) \tag{2.4}$$

Let x_i and y_i be, respectively, the binary predicted label and ground truth label for a voxel $i \in [1, n]$, where n stands for the number of voxels, and let l be the class label with $l \in \{0, 1\}$, The binary dice is defined as:

$$L_{dice} = 2\frac{\sum_{l=0}^{1} \sum_{i=1}^{n} x_{il} y_{il}}{\sum_{l=0}^{1} \sum_{i=1}^{n} (x_{il} + y_{il})} \tag{2.5}$$

Let x and y be, respectively, the class prediction and the label of a sample; the classification categorical cross-entropy is computed as follows:

$$L_{class} = -\sum_{i=1}^{n} (y \log(x) + (1 - y) \log(1 - x)) \tag{2.6}$$

We should emphasize that the classification cross-entropy loss is computed for every sample, while the segmentation cross-entropy is calculated considering each pixel within every sample. During the first phase, the cross-entropy is used as classification loss, and a weighted sum of a segmentation cross-entropy loss and the dice loss produce the segmentation loss. Then, during the second phase of 600 epochs, we only use the classification cross-entropy to train the network. We named the weight of the classification cross-entropy w_{class}, the weight of the segmentation cross-entropy loss w_{ce} and the term of the dice loss w_{dice}.

$$L_{total} = w_{class} \cdot L_{class} + w_{dice} \cdot L_{dice} + w_{ce} \cdot L_{ce} \tag{2.7}$$

Concerning the dice coefficient, we selected the following values: 0.002, 0.01, 0.02, 0.03, 0.5 and 1. The selected value impacts the gradient magnitudes during the training. For the dice coefficient of 0.02, we fixed a coefficient for the segmentation cross-entropy to half its value, in order to evaluate the possible improvements with this additional loss. We fixed the $w_{ce} = 0.01$ coefficient to be lower than the dice coefficient $w_{dice} = 0.02$. We chose these coefficients with respect to the observed values of the cross-entropy loss within the first training epochs. In order to improve the regularization of the network, we added a semi-supervised loss function that forces the model to produce low-entropy predictions and thus increases the confidence in the predictions of the classifier. We added the w_{ssl} to the global loss and fixed its value to 1.

$$L_{total} = w_{class} \cdot L_{class} + w_{dice} \cdot L_{dice} + w_{ce} \cdot L_{ce} + w_{ssl} \cdot L_E \tag{2.8}$$

2.3 Simulation Results

The training of deep learning models on the EDITH database is separated into four parts: The PE prediction based on a fusion of clinical data and images, a test of various changes in the components of a deep learning classification model, trained the model to detect PE, and finally the fine-tuning of a pretrained model. Additionally, we discuss the prediction of the risk of recurrence.

2.3.1 Database Preprocessing

The database contains information about the diagnosis of PE and the occurrence of a recurrent VTE episode. Notably, segmentation masks for the blood clot are not available; sometimes, thrombi are bounded with two crosses. We developed an algorithm to extract automatically a segmentation from these crosses. The segmentation is performed using thresholding followed by a dilation to suppress

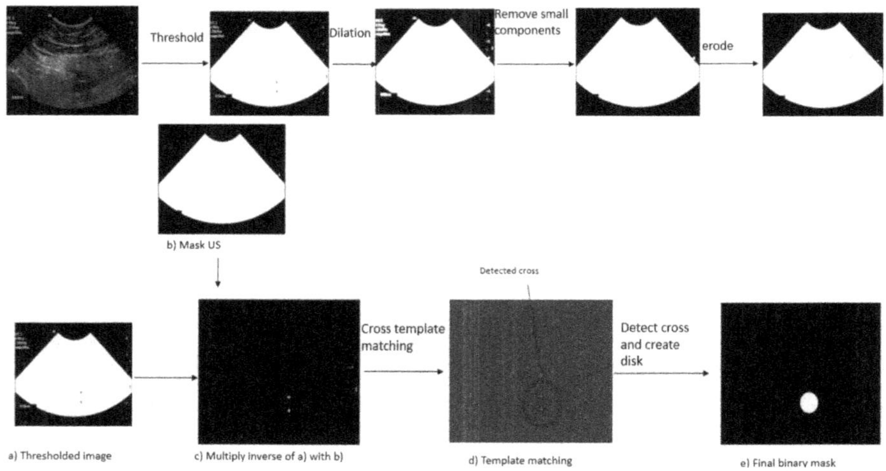

Fig. 2.4 Flowchart of segmentation of the blood clot, from the segmented ultrasound image and the binary thresholded image to the cross template matching on the cross and the extraction of the final segmentation

small components and fusion of small connecting ones, and then the crosses are restored by an erosion operator. The steps of this algorithm are illustrated in Fig. 2.4.

Theoretically, with a deep learning model using pooling layers, the size of the input image should be related to the size of the feature maps at the bottleneck of the network and the receptive field. The bottleneck of the network is situated just after the last max-pooling layer, where the feature maps are at the lowest resolution. The maximum-pooling operator of size 2×2 applied with a stride of 2 divides the size of the features by 2 at each double convolutional block. If the network contains four down-samplings and the image size is 224, the size of the feature maps at the bottleneck will be 14 (i.e $224/2^4$). The image size should be a multiple of 2^k, where k is the number of max-pooling layers, and the size of the feature maps at the bottleneck should be at least superior to the kernel size of the convolutional layer. The receptive field denotes the number of pixels in the input image that are considered by a pixel in the feature maps. The preprocessing of the segmentation model consists of centering the image on the disk and cropping various sizes centered on the center of the disk.

2.3.2 Segmentation

The segmentation of a thrombus in ultrasound imaging presents a challenging task due to the potential confusion between veins, where the blood clot occurs, and arteries. Furthermore, the detection of the clot involves visualizing it over time through several frames, along with a physical compression of the clot. Consequently, it becomes challenging for an expert to accurately recognize the clot using only a single image screened during a clinical examination. Two different preprocessing methods are applied for the segmentation task: either the entire image is used as input to the model with a down-sampling or a patch is centered on the disk created during the annotation process. We refer to these settings as the "full" or "box" images.

The model trained on the "full" image preprocessing produced too many false positives and very low dice values; thus, we focus on presenting only the results obtained on "box" images. Indeed, the deep learning segmentation model on "full" images yielded a mean dice of 0.62 with a high standard deviation of 0.31, indicating numerous segmentation failures. An example is illustrated in Fig. 2.5, where white contours represent the predicted segmentations. Notably, a significant false positive is observed on the left of the image. To address this issue, we investigated segmentations on centered "box" images using a joint classification and segmentation model. The one-phase model was trained on both tasks simultaneously throughout the entire training process. The two-phase class model was initialized on both tasks and then specifically trained for classification. Similarly, the two-phase segmentation model was initialized on both tasks and then specifically trained for the segmentation task.

Table 2.1 presents the mean and standard deviation of the dice score on the test database over "box" images centered on the clot with a disk generated as ground truth. In the binary mask, the proportion of positive pixels is high, causing less impact on the dice for false positive segmented pixels. To facilitate the segmentation task, centered processing was also used in the setting of joint classification and segmentation to reduce available contextual information.

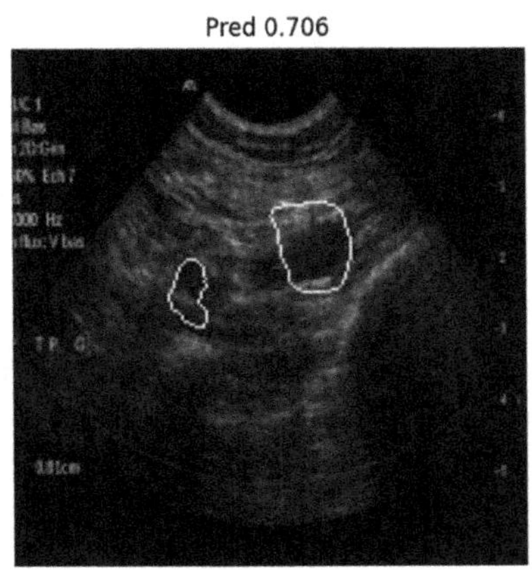

Fig. 2.5 Failed segmentation on the "full" image preprocessing with a false positive contoured with a white line on the left with Udnet; the dice value was 0.71 in this case

Table 2.1 Thrombus segmentation results on "box" images using the multi-task model

Scheme	Mean dice	Std dice	Accuracy
1-phase	0.90	0.11	0.54
2-phase segmentation	0.90	0.11	0.54
2-phase class	0.69	0.12	0.65

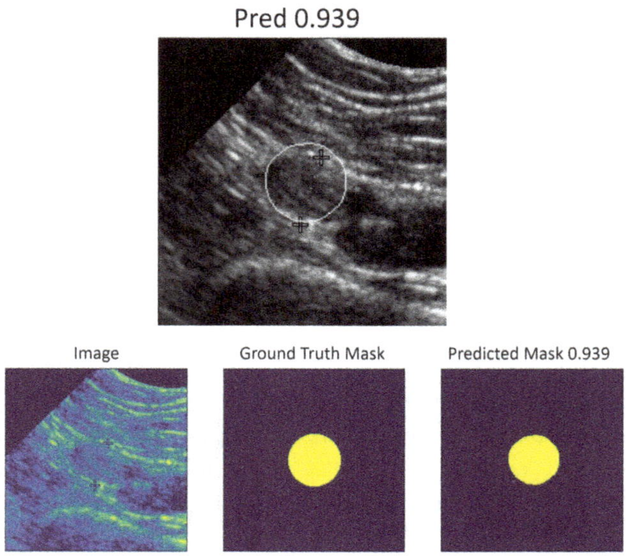

Fig. 2.6 Top Figure: Segmentation predicted by the model represented by a white contour. Predictions performed with the best segmentation model

We observe that the two-phase segmentation model produced a dice of 0.90 with a standard deviation of 0.11 but a low accuracy of 0.54. The model using only one phase produced a good mean dice of 0.90 but a low classification accuracy of 0.54. However, the two-phase model, trained specifically on classification after 200 epochs, seems to have significantly decreased the quality of the segmentation, reducing the dice by 0.34 compared to the model with one phase. With this one-phase model, accuracy increased to 0.62. It appears that the optimization of the classification score pushed the model to change many features necessary for the segmentation. The two-phase model with the low coefficient for the segmentation loss produced the best accuracy but with a mean dice score of only 0.69. Figure 2.6 displays the predicted segmentation as binary masks. We observe that the segmentation over-fitted the ground truth by drawing a disk.

The results obtained for the segmentation task on ultrasound imaging of DVT showed a tendency for the models to produce false positives, especially on entire images. The model produced better segmentations on the oversimplified task based on images centered on the annotations. Overall, the quality of the segmentations obtained seems to be insufficient for the extraction of handcrafted features.

2.3.3 Detection of PE with Handcrafted Features

Our approach is twofold. Initially, we focus on subsets of the database containing annotations on "box" images. In the second approach, we work on the entire database, employing a coarse region of interest defined randomly as a large crop centered on the original, which may or may not include the thrombus. The mean decrease impurity (MDI) feature selection [1, 14] produced the best results by selecting 34 features for the PE detection. In these features, we find 14 gray-level run length matrix (GLRLM) features [8], 6 gray-level dependence matrix (GLDM) features [5, 24], 6 gray-level dependence matrix (GLDM) features [21], 4 gray-level size zone matrix (GLSZM) features [22] and five first-order features.

Table 2.2 presents obtained results with a standard image normalization, centered around the annotation with a crop of size (144, 144). We observed similar results without image normalization; the best accuracy, 0.74, was obtained with the random forest and a standard normalization of features. This accuracy is 0.02 higher than the same model without image normalization. This model (random forest and a standard normalization of features) also shows a balanced sensitivity and specificity with a score of 0.72 and 0.82, respectively. The best sensitivity in these results with a standard image normalization is provided by the regression and a minmax normalization, with a sensitivity of 0.78 and an accuracy of 0.68. Models with a standard image normalization increase the mean accuracy by 0.05, the mean sensitivity by 0.02 and the mean specificity by 0.07.

Table 2.2 Performance metrics on various classifiers and handcrafted feature normalization with a standard image normalization: accuracy (ACC), F1-score (F1), sensitivity (Sensi) and specificity (Speci)

	RMSE	ACC	F1	Sensi	Speci
Random forest					
No normalization	0.56	0.67	0.59	0.59	0.81
Standard normalization	0.51	0.74	0.71	0.72	0.82
minmax normalization	0.56	0.68	0.62	0.60	0.81
PCA (feature reduction)	0.60	0.63	0.44	0.38	0.77
SVM					
No normalization	0.73	0.46	0.32	0.38	0.60
Standard normalization	0.67	0.54	0.58	0.68	0.52
minmax normalization	0.56	0.69	0.67	0.73	0.75
PCA (feature reduction)	0.64	0.58	0.62	0.77	0.55
Regression					
No normalization	0.72	0.48	0.56	0.77	0.26
Standard normalization	0.66	0.57	0.58	0.67	0.58
minmax normalization	0.56	0.68	0.69	0.78	0.71
PCA (feature reduction)	0.60	0.63	0.66	0.78	0.62

2.3.4 Detection of Pulmonary Embolism with Deep Learning Models

This section presents our results on the fusion of image and clinical data, then the modification of various components of the deep neural network for the classification based solely on images. We modify the depth (number of convolutional layers), width (number of filters) of the model as well as the optimizers, normalizations and activation functions. Then, we test the transfer learning of a pretrained vision transformer model on our classification task. The results are computed using cross-validation across various splits of the database into training, validation, and testing sets. A split or a fold corresponds to a single partition of the database into the training, validation and test sets. Multiple runs for the same hyperparameter set are performed on each fold. However, consistent results are not always achieved for identical hyperparameters across different runs, primarily due to the randomness associated with the non-fixed selection of data augmentation operations for each batch. The observed differences can be significant and may also be related to changes in the optimization process.

2.3.4.1 Clinical Data Fusion

To evaluate the performance of both the fusion model and the image model, we divided our database into three subsets (DB1, DB2 and DB3) depending on the collected protocol (including the medical equipment and the provided hospital). Then, we train first the models on all three databases. Subsequently, we train both models on only DB1 and DB2, and, finally, on DB1 only and DB2 only. Notably, we do not conduct training and testing on DB3 alone due to its relatively small size. The models are trained until there is no improvement observed on the validation set for 120 epochs and up to 800 epochs. At each batch, two data augmentation operations are performed randomly among left or right flip, cropping, sharpening, affine transformation, linear contrast, Gaussian blur, additive Gaussian noise, edge detection, dropout and elastic transformation, using imgaug package [10]: the flip consists in mirroring the image, the crop is a selection of a portion of the image, the sharpening increases the contrast at the edges, the affine transformation performs translation rotation or scaling and the linear contrast modifies the contrast of the image linearly. The additive Gaussian noise consists in adding Gaussian noise, proportional to the variance of pixels in the image. The edge detection applies a canny filter, which consists of a smoothing gradient filter and edge tracking with hysteresis thresholding. The learning rate is set to 10^{-5}, the number of filters to 32, the batch size to 4. On this hyperparameter set, we evaluate two different depths (three or four poolings) for both the fusion model and the image-based model. Here, modifying the depth of the network involves adjusting the number of double convolutional blocks and the corresponding number of max-pooling layers in between. The three-pooling model consists in using three-max-pooling between

Table 2.3 Accuracy on subsets with fusion or image-only models with three or four pooling layers

Database	Mode	Down-samplings	Accuracy
DB 1+2+3	Fusion	4	0.61
DB 1+2+3	Fusion	3	0.62
DB 1+2+3	Image	4	0.63
DB 1+2+3	Image	3	0.6

four-double convolutional blocks. The model contains 1.2 millions of trainable parameters. The four-pooling model consists in using four-max-pooling between five-double convolutional blocks. This model contains 2.3 million parameters, and the total memory size for an input of 0.19 megabytes is estimated to be 189.5 megabytes.

The best eightfold accuracy of the PE occurrence classification using all databases with an image-only model and four down-samplings was 0.63 (see Table 2.3). The accuracy of the fusion model was slightly lower with a decrease of 0.02 with the four down-sampling model. However, the fusion model was 0.02 higher than the image-only model, with the three down-sampling architecture.

The experiments on the fusion model with a concatenation finally did not show improvements with the addition of clinical data, expect for the three down-samplings model on DB1 and DB2. Additionally, the experiments with the metablock on DB1 and DB2 provided an accuracy of 0.58, which is lower than the model based on a concatenation. We conclude that either the clinical data did not provide enough information to increase the accuracy, or the concatenation and the metablock model did not allow leverage of this clinical data.

2.3.4.2 Variation of the Model Width and Depth on DB1

We tested several variations on our classification model and hyperparameters for the detection of PE with a model based solely on images without clinical data. We conducted all experiments on DB1. Our modifications on the optimizer, activation function or normalization reflect the adjustments implemented in the ConvNext study [12]. We restrain the scope of our study to a width of either 32 or 64 filters in the initial layer, systematically doubling the number of filters at each successive double convolutional block. We compare the accuracy obtained with both settings on the model with two, three or four poolings. On the two-pooling model, an improvement of 5.5% was observed using 64 filters compared to 32 filters (see Table 2.4).

Table 2.4 Impact on accuracy using 64 filters instead of 32 filters at different model depths

Model depth	Change in fold-mean accuracy with doubling filters
4-poolings	Decrease: 15.6%
3-poolings	Decrease: 6.4%
2-poolings	Increase: 5.5%

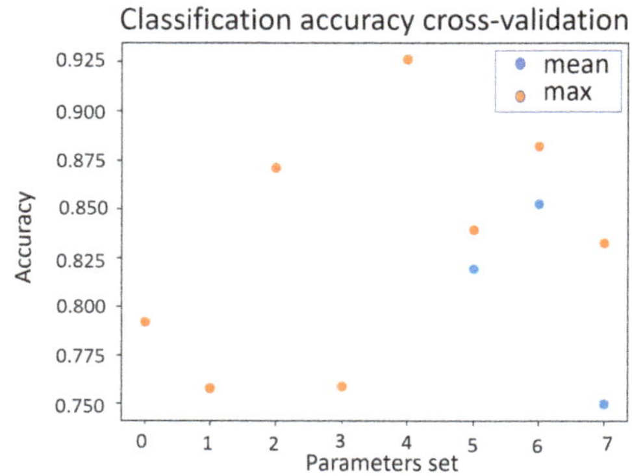

Fig. 2.7 Accuracy fold-mean and fold-max over all splits in cross-validation with the best hyperparameter set

On the four-pooling model, there is a 15.6% improvement using the 32 filter base instead of using the 64 filter base. On the three-pooling model, there is a 6.4% improvement using the 32 filter base instead of the 64 filter base. This trend highlights that as we increase the depth of the network, augmenting the width tends to degrade performance. Nevertheless, it is noteworthy that the model with the greatest depth still provides the best accuracy. We observe when using the training loss curve that a wider model might lead to over-fitting. Based on the results shown in Table 2.4, we established the foundational hyperparameter values for subsequent experiments. We systematically varied each of these hyperparameters to assess their impact on model metrics.

The number of filters set at 32 influences feature extraction, while the fixed batch size of 4 impacts training stability, constrained by memory limitations. The number of augmentations, initially set at 2, enhances model generalization and undergoes testing at various values in our experiments. The image shape (224, 224) and the crop of 448 define input dimensions, influencing feature complexity and size at each layer. The activation functions, based on rectified linear unit (ReLU), offer computational efficiency and nonlinearity for positive values, promoting smooth optimization. The learning rate, set at 10^{-5}, governs optimization step sizes and consistently delivers optimal results in our tasks.

Our study aims to establish a connection between the texture and aspect of the blood clot and the PE detection, with a broader focus than sensitivity or specificity. However, we provide all metrics for the best model based on the accuracy. Figure 2.7 provides the fold-mean and fold-max accuracy for all the splits in cross-validation for the best hyperparameter set. We can see that the accuracy reaches the maximum value near 0.925 and a minimum value near 0.75. In a few cases, the fold-mean and fold-max values were similar.

2.3.4.3 Various Kernel Sizes and Learning Rates for PE Detection

The subsequent tests are conducted using various kernel sizes for the convolutions. Increasing the kernel size allows the capture of more complex and global features. Additionally, it enlarges the receptive field of the model, which denotes the number

Table 2.5 Various learning rates on cross-validation

Learning rate	Fold-mean accuracy	Fold-max accuracy
10^{-5}	0.78	0.83
10^{-4}	0.63	0.86
$5 * 10^{-6}$	0.77	0.77

of pixels from preceding layers that influence a specific pixel in a subsequent layer. However, increasing the kernel size increases the number of parameters and thus increases both the computational cost and the risk of over-fitting.

In Table 2.5, we observe that the learning rate of 10^{-5} offers the best consistency with a fold-mean accuracy, where the best fold-max accuracy was observed with 10^{-4}. With the learning rate of 10^{-5}, we obtained a fold-mean accuracy of 0.83 on cross-validation with a standard deviation of 0.06. Additionally, a sensitivity of 0.86 was obtained with a standard deviation of 0.1 and a specificity of 0.69 with a standard deviation of 0.22.

2.3.4.4 Various Activation Functions and Normalizations for PE Detection

The replacement of all ReLU activations in the model by the Gaussian error linear unit (GELU) activation provided a fold-max accuracy of 0.82 and a fold-mean accuracy of 0.75 using a base of 32 filters in the first layer and using a learning rate of 10^{-5}. Various parameters such as the filter number (32 or 64) and the learning rates were tested. Our results did not show any improvement in the accuracy using the GELU. This is contradictory with other results in the literature [12], who proved that GELU provided better performance on the image classification task. The correction of the vanishing or exploding gradients provided by GELU might be more important on larger models trained on large-scale datasets. This would explain why it did not improve the model's performance in our case (Table 2.6).

After the study of activation functions, we study the impact of replacing the batch normalization (BN) with a layer normalization (LN). We named LN1 the layer normalization applied to only one block of each double convolutional block. The layer normalization is denoted by (LN) and applied on all activation functions used in the network. The results are similar with the LN1 and LN approaches and both provide a lower accuracy than when using the batch normalization (Table 2.7).

Table 2.6 Accuracy obtained with GELU on cross-validation

Learning rate	Filters	Fold-mean accuracy	Fold-max accuracy
10^{-5}	32	0.75	0.82
10^{-5}	64	0.64	0.66
$5 * 10^{-5}$	64	0.58	0.69

Table 2.7 Accuracy obtained when replacing BN by LN or LN1 on cross-validation

Learning rate	Normalization	Fold-mean accuracy	Fold-max accuracy
10^{-5}	LN	0.75	0.82
10^{-5}	LN1	0.64	0.66

Table 2.8 Eightfold accuracy when with the ranger21 optimizer

Learning rate	Optimizer	Fold-mean accuracy	Fold-max accuracy
10^{-4}	Adam	0.63	0.86
10^{-5}	Adam	0.78	0.83
10^{-4}	Ranger21	0.71	0.84
10^{-5}	Ranger21	0.69	0.86
10^{-5}	Lion	0.59	0.73

2.3.4.5 Various Optimizers and Test of Fransfer Learning for PE Detection

The training with the lion optimizer provided 10% lower accuracy than the ones obtained with the Adam optimizer, with a batch size of 16 and a learning rate of 10^{-5}. However, the authors of [3] advise a large batch size of 4096 to leverage the use of the lion optimizer which is inapplicable in our project, as our data contains less than 4096 images. In our experimental results, the lion optimizer provided both the lowest fold-mean and fold-max accuracy, and the best performance, with cross-validation, was obtained with the Adam optimizer and provided an accuracy of 0.83 with a sensitivity of 0.69 and a specificity of 0.86. Additionally, we used the fine-tuning of the linear layer of a pretrained VIT16 model [6]. However, we observed an accuracy around 0.72. Table 2.8 shows the results with the ranger21 optimizer [25].

2.3.5 Classification of Recurrent (VTE) with Deep Learning Models

The training of the four-pooling and five-pooling models did not provide conclusive results. As the accuracy is not representative for an unbalanced dataset with only a few positive cases, we considered the F1-score as well. Across all tests, we achieved a fold-max F1-score of 0.22, indicating a notably low performance. Due to these consistently low scores, we are unable to draw positive conclusions regarding the prediction of recurrence in ultrasound imaging. However, we observed on one split of the dataset a specificity of 0.82 with an accuracy of 0.74 and a sensitivity of 0.5. However, the number of positive cases is too small to train a robust model.

2.4 Conclusion

The first experiments performed with the deep learning model, on the raw database, were non-conclusive and the optimization did not start at all. We call raw database the data gathered from clinical examinations, where the clinicians screened and saved several images during the examination. We then performed a review of the database with a program allowing the experts to select the relevant ultrasound images of proximal DVT. The experiments on the classification of the occurrence of a PE (or the detection of PE), with handcrafted features and machine learning models provided the best accuracy of 0.74 with a sensitivity of 0.72 and a specificity of 0.82. However, this was applied on a subset of the dataset in which annotations were available indicating the position of the thrombus. The model was based on 34 selected features. The results provide a better score than the accuracy of 0.66 obtained in the previous project based on the scattering operator.

In terms of methodology, first, we developed architectures of deep neural networks for classification and segmentation that can be common to both tasks. Second, we developed handcrafted features and selected them specifically to each task. We should notice that in both tasks, we face the same challenges related to the ultrasound imaging modality (noise and artifacts).

Concerning the detection of PE, the combination of handcrafted features with machine learning models achieved a maximum accuracy of 0.74, whereas the deep learning model achieved a higher accuracy of 0.86. Despite the deep learning models having over 1 billion parameters, their inference on a single image is faster than 300 ms using only a simple CPU. The development of machine learning models and the use of handcrafted features offer additional insights into the classification process but require an initial segmentation of the blood clot. Furthermore, the resulting classification model based on handcrafted features remains complex and challenging to interpret.

Acknowledgments This research project is supported by the French Clinical Research Infrastructure Network on Venous Thrombo-Embolism (FCRIN INNOVTE). The authors would also like to acknowledge Brest University Hospital and ENSTA Bretagne for their support.

References

1. Breiman, L.: Classification and Regression Trees. Routledge, England (2017)
2. Cao, H., Wang, Y., Chen, J., Jiang, D., Zhang, X., Tian, Q., Wang, M.: Swin-unet: Unet-like pure transformer for medical image segmentation. In: European Conference on Computer Vision, pp. 205–218. Springer, Berlin (2022)
3. Chen, X., Liang, C., Huang, D., Real, E., Wang, K., Liu, Y.,, Pham, H., Dong, X., Luong, T., Hsieh, C.-J., et al.: Symbolic discovery of optimization algorithms. arXiv preprint arXiv:2302.06675 (2023)

4. Chorowski, J.K., Bahdanau, D., Serdyuk, D., Cho, K., Bengio, Y.: Attention-based models for speech recognition. In Cortes, C., Lawrence, N., Lee, D., Sugiyama, M., Garnett, R. (eds.). Advances in Neural Information Processing Systems, vol. 28. Curran Associates, Inc., New York (2015)
5. Conners, R.W., Harlow, C.A.: A theoretical comparison of texture algorithms. IEEE Trans. Pattern Anal. Mach. Intell. **3**, 204–222 (1980)
6. Dosovitskiy, A., Beyer, L., Kolesnikov, A., Weissenborn, D., Zhai, X., Unterthiner, T., Dehghani, M., Minderer, M., Heigold, G., Gelly, S., et al.: An image is worth 16 × 16 words: Transformers for image recognition at scale. arXiv preprint arXiv:2010.11929 (2020)
7. Duffett, L., Castellucci, L.A., Forgie, M.A.: Pulmonary embolism: update on management and controversies. BMJ **370**, m2177 (2020)
8. Haralick, R.M., Shanmugam, K., Dinstein, I.: Textural features for image classification. IEEE Trans. Syst. Man Cybern. **SMC-3**(6), 610–621 (1973)
9. Isensee, F., Jaeger, P.F., Kohl, S.A.A., Petersen, J., Maier-Hein, K.H.: nnU-Net: a self-configuring method for deep learning-based biomedical image segmentation. Nat. Methods **18**(2), 203–211 (2021)
10. Jung, A.B., Wada, K., Crall, J., Tanaka, S., Graving, J., Reinders, C., Yadav, S., Banerjee, J., Vecsei, G., Kraft, A., Rui, Z., Borovec, J., Vallentin, C., Zhydenko, S., Pfeiffer, K., Cook, B., Fernández, I., De Rainville, F.-M., Weng, C.-H., Ayala-Acevedo, A., Meudec, R., Laporte, M., et al.: imgaug. https://github.com/aleju/imgaug (2020). Online; accessed 01-Feb-2020
11. Kingma, D.P., Ba, J.: Adam: A method for stochastic optimization. In: Bengio, Y., LeCun, Y. (eds.) 3rd International Conference on Learning Representations, ICLR 2015, San Diego, CA, USA, May 7–9, 2015, Conference Track Proceedings (2015)
12. Liu, Z., Mao, H., Wu, C.-Y., Feichtenhofer, C., Darrell, T., Xie, S.: A convnet for the 2020s. In: Proceedings of the IEEE/CVF Conference on Computer Vision and Pattern Recognition, pp. 11976–11986 (2022)
13. Long, J., Shelhamer, E., Darrell, T.: Fully convolutional networks for semantic segmentation. In: Proceedings of the IEEE Conference on Computer Vision and Pattern Recognition, pp. 3431–3440 (2015)
14. Louppe, G., Wehenkel, L., Sutera, A., Geurts, P.: Understanding variable importances in forests of randomized trees. Adv. Neural Inf. Proces. Syst. **26**, 1–9 (2013)
15. Oktay, O., Schlemper, J., Folgoc, L.L., Lee, M., Heinrich, M., Misawa, K., Mori, K., McDonagh, S., Hammerla, N.Y., Kainz, B., et al.: Attention u-net: Learning where to look for the pancreas. arXiv preprint arXiv:1804.03999 (2018)
16. Pacheco, A.G.C., Krohling, R.A.: An attention-based mechanism to combine images and metadata in deep learning models applied to skin cancer classification. IEEE J. Biomed. Health Inform. **25**(9), 3554–3563 (2021)
17. Parikh, A.P., Täckström, O., Das, D., Uszkoreit, J.: A decomposable attention model for natural language inference. arXiv preprint arXiv:1606.01933 (2016)
18. Paszke, A., Gross, S., Massa, F., Lerer, A., Bradbury, J., Chanan, G., Killeen, T., Lin, Z., Gimelshein, N., Antiga, L., Desmaison, A., Kopf, A., Yang, E., DeVito, Z., Raison, M., Tejani, A., Chilamkurthy, S., Steiner, B., Fang, L., Bai, J., Chintala, S.: Pytorch: An imperative style, high-performance deep learning library. In: Wallach, H., Larochelle, H., Beygelzimer, A., d'Alché-Buc, F., Fox, E., Garnett, R. (eds.) Advances in Neural Information Processing Systems, vol. 32, pp. 8024–8035. Curran Associates, Inc., New York (2019)
19. Ronneberger, O., Fischer, P., Brox, T.: U-net: Convolutional networks for biomedical image segmentation. In: International Conference on Medical Image Computing and Computer-Assisted Intervention, pp. 234–241. Springer, Berlin (2015)
20. Sandler, D.A., 1292084 Martin, J.F.: Autopsy proven pulmonary embolism in hospital patients: are we detecting enough deep vein thrombosis? J. R. Soc. Med. **82**(4), 203–205 (1989)
21. Sun, C., Wee, W.G.: Neighboring gray level dependence matrix for texture classification. Comp. Vision, Graphics, Image Process. **23**(3), 341–352 (1983)
22. Thibault, G., Angulo, J., Meyer, F.: Advanced statistical matrices for texture characterization: application to cell classification. IEEE Trans. Biomed. Eng. **61**(3), 630–637 (2013)

23. Vaswani, A., Shazeer, N., Parmar, N., Uszkoreit, J., Jones, L., Gomez, A.N., Kaiser, Ł., Polosukhin, I.: Attention is all you need. Adv. Neural Inf. Proces. Syst. **30**, 1–15 (2017)
24. Weszka, J.S., Dyer, C.R., Rosenfeld, A.: A comparative study of texture measures for terrain classification. IEEE Trans. Syst. Man Cybern. **4**, 269–285 (1976)
25. Wright, L., Demeure, N.: Ranger21: a synergistic deep learning optimizer. arXiv preprint arXiv:2106.13731 (2021)
26. Zhang, Q., Cui, Z., Niu, X., Geng, S., Qiao, Y.: Image segmentation with pyramid dilated convolution based on resnet and u-net. In: Neural Information Processing: 24th International Conference, ICONIP 2017, Guangzhou, China, November 14–18, 2017, Proceedings, Part II 24, pp. 364–372. Springer, Berlin (2017)

Chapter 3
Synchronization-Driven Community Detection: Dynamic Frequency Tuning Approach

Abdelmalik Moujahid and Alejandro Cervantes Rovira

Abstract Many real-world networks, spanning social, communication, and biological domains, exhibit temporal dynamics in which relationships between nodes evolve over time. In these dynamic networks, communities are not static entities but are subject to continuous changes in their structure, composition, and interaction over time. Conventional community detection algorithms, which typically analyze static snapshots of networks, often fail to capture the underlying dynamics, leading to an incomplete understanding of network organization. Therefore, there is growing interest in developing algorithms that are able to recognize communities in dynamic networks, taking into account the temporal evolution of node memberships and community structures. Dynamic community detection algorithms typically work with sequences of time frames, where each frame represents the network structure at a particular point in time. These algorithms aim to dynamically update network communities by utilizing information from previous time frames. In this context, synchronization-based algorithms represent a promising approach. By exploiting the emerging synchronization patterns within the network, these algorithms identify communities of closely connected nodes, often corresponding to communities or clusters. In particular, we focus on an algorithm that incorporates dynamic frequency tuning mechanisms that allow for evolving network dynamics and improve the accuracy of community detection over time.

Keywords Complex systems · Community detection · Clustering · Dynamical systems · Synchronization

3.1 Introduction

The organizational pattern of communities is a widespread phenomenon evident across various networked systems spanning biological, social, technological, and

A. Moujahid (✉) · A. C. Rovira
Universidad Internacional de la Rioja (UNIR), Logroño, Spain
e-mail: abdelmalik.moujahid@unir.net; alejandro.cervantesrovira@unir.net

F. Dornaika et al. (eds.), *Advances in Data Clustering*,
https://doi.org/10.1007/978-981-97-7679-5_3

information domains [22]. Within the framework of network science, a network is depicted as a graph $\mathcal{G}$, consisting of numerous interconnected nodes or vertices. A community within such a network is distinguished by dense internal connections among its constituent nodes, contrasting with sparse connections linking nodes across different communities [11].

The goal of detecting communities in graphs is to uncover these connected modules and possibly explain their hierarchical organization based solely on the structural properties encoded in the topology of the graph. This problem has historical roots and has manifested itself in various forms in numerous disciplines [6, 9, 13, 16]. In particular, the fundamental concept of community structure in complex networks was originally emphasized by Girvan and Newman in the physics literature [11].

Over the years, a variety of algorithms have been developed to address the challenge of community detection. These range from traditional methods such as graph partitioning and spectral clustering to more modern approaches such as modularity-based techniques and synchronization-based dynamic algorithms [7, 15, 27]. For a comprehensive overview of the current state of the art and methodology in community detection, including recent advances and emerging trends, we refer the reader to Fortunato's report [10].

Dynamic algorithms based on synchronization have attracted a lot of attention in research, as demonstrated by the works of Arenas et al. [3], Boccaletti et al. [5], Li et al. [19], and Almendral et al. [2]. These studies have shown that exploiting the dynamic synchronization process in complex networks is a promising approach to reveal their different topological scales. Many of these studies are based on the Kuramoto model [17] or its variants, such as that of Pluchino et al. [1], in which the dynamics of each individual oscillator is encapsulated in a one-dimensional state space.

However, real systems usually exhibit high-dimensional, complex dynamics characterized by numerous variables and parameters. Consequently, there is a practical interest in understanding the community structure in complex networks under such circumstances. Pluchino et al. [26] addressed this issue by considering N identical Rössler oscillators coupled according to the dynamics of the opinion change rate model. They used the edge betweenness load matrix, albeit at the cost of additional computational complexity. Nevertheless, it is important to point out that these algorithms often depend on certain initial conditions and frequency distributions of the coupled oscillators, which leads to the appearance of unstable solutions.

In this study, we follow an approach in which we represent the dynamics of a complex network by an ensemble of nonidentical coupled chaotic systems, where each system has a certain natural frequency. The use of nonidentical coupled systems offers several advantages for modeling the dynamics of complex networks. First, many complex networks generally consist of nonidentical entities with different properties and behaviors. For example, in biological networks, the nodes may represent different types of cells or organisms with different genetic compositions or physiological properties. Similarly, individuals in social networks have different characteristics, interests, and behaviors. Ignoring this heterogeneity by assuming identical nodes oversimplifies reality and can lead to inaccurate predictions and limited insights into network dynamics.

The chapter is organized as follows: Sect. 3.2 introduces the key aspects of the community detection approach. In Sect. 3.3, the dynamics of the oscillator network is discussed. Section 3.4 provides a comprehensive explanation of the dynamic tuning approach utilized for adjusting the characteristic frequencies of coupled oscillators, alongside outlining the procedure for generating dynamic graphs from time series data and delving into the modularity maximization approach. Section 3.5 presents an overview of the experimental setup, describing the real networks used in this work (Sect. 3.5.1) and providing insight into the dynamics of the Rössler coupled oscillators (Sect. 3.5.2). Section 3.6 presents simulation results derived from both social and computer-generated networks to assess the efficacy of our methodology. Finally, conclusions are drawn in Sect. 3.7.

3.2 Modeling Community Structures in Networked Systems

In this study, we describe an approach aimed at discovering the community structures embedded within networked systems. The methodology relies on an ensemble of nonidentical chaotic systems governed by a predefined topology. By emulating the dynamic interplay observed in real-world social networks, we focus on the evolving nature of interactions, reflecting the gradual intensification of connections over time. Through this modeling framework, we seek to capture the emergence and consolidation of diverse communities within the network, thereby shedding light on its underlying organizational principles.

First, we use an ensemble of interacting, nonidentical chaotic systems where the strength of coupling increases linearly over time. This setup simulates a social scenario in which the interactions between agents become more and more intense. In particular, this increase in connections within the society facilitates the stabilization of several different, noninteracting communities [1].

Let $\mathcal{G}$ represent a graph with adjacency matrix A and Laplacian matrix $\mathcal{L}$. The Laplacian matrix $\mathcal{L}$ is defined as $\mathcal{L} = D - A$, where D is the diagonal degree matrix. The strength of the coupling, denoted as σ, spans a spectrum from zero to a maximum value determined by the ratio:

$$\sigma_{\max} = \lambda_{\max}/\lambda_2, \tag{3.1}$$

where λ_2 and $\lambda_{\max}$ represent the first nonzero and maximum eigenvalues of the Laplacian matrix $\mathcal{L}$, respectively. This ratio represents an intermediate state characterized by highly interconnected units that form localized, synchronized communities within the network. The value of σ ranges from 0, indicating an uncorrelated initial state, to the maximum value determined by the eigenvalue ratio.

The relationship between the spectral properties of the Laplacian matrix and the hierarchical emergence of communities has been shown in previous research [3].

This connection emphasizes the importance of spectral information in revealing the underlying community structure in complex networks. By taking advantage of this relationship, our approach aims to exploit the evolving spectral properties of the Laplacian matrix to reveal the hierarchical organization of communities with increasing coupling strength over time. Using this methodology, we seek to uncover the intricate interplay between network dynamics and community formation and to elucidate the underlying mechanisms that govern the structure and evolution of complex systems.

The second aspect of this approach refers to the influence that neighbors exert on the change in natural frequency (or opinion) of individual oscillators (or agents), thus contributing to the emergence of the modular structure observed in the network. In this context, each module corresponds to oscillators that share a common average frequency. Without coupling, each oscillator initially evolves autonomously according to its natural frequency, leading to uncorrelated states. However, as soon as coupling is activated, new correlated states emerge corresponding to the oscillators evolving at the same average frequency. We show that the dynamic adaptation of the characteristic frequencies of the oscillators, controlled by the concept of confidence [14], increases the stability of these emerging correlated states. Consequently, this dynamic adaptation process leads to a final frequency vector that emphasizes the community structure inherent in the network. To achieve this goal, we follow a parameterless frequency adaptation approach. By integrating this mechanism into this methodology, we aim to elucidate how the dynamic adjustment of oscillator frequencies in response to neighbor influences contributes to the delineation of network communities. This approach not only improves the stability of emergent states but also facilitates the identification of cohesive modules within the network, thus providing valuable insights into the underlying principles of community formation in complex systems.

3.3 Network Dynamics

The dynamics of a graph G comprising N nonidentical coupled oscillators can be effectively described by a system of differential equations. Each oscillator's behavior, indexed by i, is governed by the equation:

$$\dot{\mathbf{x}}_i(t) = \mathbf{f}(\mathbf{x}_i, \omega_i) + \frac{\sigma}{k_i} \sum_{j=1}^{N} \mathcal{A}_{ij} \mathbf{g}(\mathbf{x}_j - \mathbf{x}_i), \tag{3.2}$$

where $\mathbf{x}_i \in \mathbb{R}^d$ represents the state vector of the ith oscillator at the time t. Here, $\dot{\mathbf{x}}_i(t) = \mathbf{f}(\mathbf{x}_i, \omega_i)$ refers to the intrinsic dynamics of the node i, which is characterized by its natural frequency ω_i.

All oscillators are nonidentical, and the parameter $\omega_i = \omega_0 + \Delta\omega_i$ selected randomly from a uniform distribution corresponds to the natural frequency of the individual oscillator. $\Delta\omega_i$ is the frequency mismatch between neighboring chaotic oscillators.

The oscillators are interconnected through a predefined network topology described by the $N \times N$ adjacency matrix ($\mathcal{A}_{ij}$), and $\mathbf{g}(\mathbf{x}_j - \mathbf{x}_i)$ is the output function through which units interact. The strength of the coupling between the ith and the jth oscillator is given by σ if they are connected in the graph; otherwise it is zero. Finally, k_i is the degree of the ith oscillator within the network.

This formulation allows for a comprehensive understanding of the collective behavior emerging from the interactions among nonidentical oscillators, incorporating both their individual dynamics and their interconnections within the network topology.

3.4 Dynamic Tuning Approach

The tuning mechanism relies on the impact neighbors have on altering the inherent frequency (or opinion) of each oscillator (or agent), thereby playing a role in the formation of the observed modular structure within the network. In this context, each module corresponds to oscillators that share a common average frequency. Without coupling, each oscillator initially evolves autonomously according to its natural frequency, leading to uncorrelated states. However, when coupling is activated, new correlated states are created in which the oscillators develop with the same average frequency.

With a coupling strength $\sigma = 0$, each oscillator evolves according to its own dynamics, which leads to uncorrelated state space variables. As soon as the coupling strength increases, several oscillators may synchronize and oscillate at a common average frequency, while their neighbors have their own different frequencies.

To enhance the stability of these emergent regimes, we propose a parameter-free dynamical adaptation mechanism to change the characteristic frequencies of oscillators. Each oscillator iteratively adjusts its characteristic frequency based on the median frequency of neighboring oscillators within a confidence interval. Oscillators whose dynamics are within this confidence limit are considered influential in determining the frequency adjustments.

The presented algorithm (see Algorithm 1) describes an adaptive frequency tuning mechanism for coupled oscillators that aims to achieve synchronized states while avoiding homogeneity of the frequency vector. It starts with the initialization of the connectivity matrix and the calculation of the maximum coupling strength based on the network topology. Through iterative steps, the algorithm solves the system of ordinary differential equations (ODEs) of the coupled oscillators, constructs similarity and dynamic connectivity matrices, and updates the oscillator frequencies based on the interactions between the neighbors. The strength of the coupling is gradually increased over the iterations to facilitate the adjustment. This approach enables dynamic adjustments to the frequencies of the oscillators and promotes synchronization and clustering within the network structure.

3.4.1 The Main Algorithm

The algorithm below (see Algorithm 1) describes the adaptive frequency tuning mechanism, as discussed earlier, aimed at achieving synchronized states while preventing uniformity in the frequency vector. It starts with the initialization of the connectivity matrix and the calculation of the maximum coupling strength based on the network topology. Through iterative steps, the algorithm solves the system of ODEs of the coupled oscillators, constructs similarity and dynamic connectivity matrices (see Sect. 3.1), and updates the oscillator frequencies based on the interactions between the neighbors. In fact, for each oscillator i, we define its neighbors of confidence, $\mathcal{N}(i)$, as the set of adjacent oscillators according to the dynamic connectivity matrix $\mathcal{A}^*$ given by Eq. 3.4. The frequency of oscillator i is then replaced by the median frequency of oscillators forming the confidence neighbors. Using the median instead of the mean frequency tends to be more robust to outliers and gives rise to a frequency vector with a clear clustering structure.

The strength of the coupling is gradually increased over the iterations to facilitate the adjustment. This approach enables dynamic adjustments to the frequencies of the oscillators and promotes synchronization and clustering within the network structure.

Algorithm 1: Adaptive frequency tuning algorithm

Data: Initial coupling strength $\sigma = 0$, adjacency matrix $\mathcal{A}$, number of iterations T, simulation time t_{sim}
Result: Vector of oscillator frequencies
Initialize the connectivity matrix to $\mathcal{A}^*$;
Initialize the frequency vector to random uniform values in the interval $[\omega_0 - \Delta\omega, \omega_0 + \Delta\omega]$;
Compute the maximum coupling strength σ_{max} according to Eq. 3.1;
Compute the adaption step as $\Delta_\sigma = \frac{\sigma max}{T}$;
for $n = 1$ **to** T **do**
    Solve ODE system of coupled oscillators given by Eq. 3.2 over a duration of t_{sim} time steps ;
    Construct the similarity matrix **S** based on the time series data, where each data point represents the state of an oscillator;
    foreach *oscillator i* **do**
        Compute average similarity m_i with all other oscillators using S according to Eq. 3.3;
        Define confidence neighbors $\mathcal{N}(i)$ based on $\mathcal{A}^*$;
        Update frequency of oscillator i using median frequency of $\mathcal{N}(i)$;
    end
    Increment the coupling strength $\sigma = \sigma + \Delta_\sigma$;
end

3.4.2 From Time Series to Similarity Graph

Based on the time series data, we build a similarity graph employing the normalized Pearson's coefficients. Let p_{ij} denote the Pearson's coefficient between two arbitrary time series y_i and y_j corresponding to the temporal evolution of the y component of oscillators i and j, respectively. These Pearson's coefficients are transformed to similarities that belong to the interval [0, 1]:

$$s_{ij} = \frac{p_{ij} - \min(p_{ij})}{1 - \min(p_{ij})}$$

where $\min(p_{ij})$ denotes the minimum p_{ij} over S's entries. Based on this similarity matrix, for each oscillator i, we can compute its average similarity to all other oscillators denoted as m_i:

$$m_i = \frac{1}{N-1} \sum_{j=1, j\neq i}^{N} s_{ij}. \tag{3.3}$$

where N is the total number of oscillators and s_{ij} represents the similarity between oscillator i and j as specified by the similarity matrix $\mathbf{S}$.

The matrix S is converted to a dynamic connectivity matrix $\mathcal{A}^*$ that will reflect the dynamics of real associations between the oscillators during the synchronization process. This matrix is constructed according to an adaptive neighborhood process as follows:

$$\mathcal{A}^*_{ij} = \begin{cases} \mathcal{A}_{ij} & \text{if } s_{ij} > m_i \text{ or } s_{ij} > m_j \\ 0 & \text{otherwise} \end{cases} \tag{3.4}$$

3.4.3 Optimal Network Partitioning

The adaptation procedure gives rise to a one-dimensional frequency vector with some clustering structure able to detect the underlying modules present in a given network. In each adaptation step, a new frequency vector is obtained, and then new community subdivisions will emerge. So, to determine the optimal number of communities or clusters, we have adopted the criterion of maximum modularity [23].

Modularity quantifies the extent to which similar entities are connected in a network by evaluating the prevalence of edges within communities compared to a random graph. Mathematically, it is calculated as the discrepancy between the observed and expected number of edges connecting vertices of similar type [24].

For a weighted graph $\mathcal{G}$ with connectivity matrix $\mathcal{A}$ and a given partition of its nodes into C communities $\{c_1, c_2, \ldots, c_K\}$, the modularity $\mathcal{M}$ can be formulated as follows [24]:

$$\mathcal{M} = \frac{1}{2\mathcal{W}} \sum_{ij} (\mathcal{A}_{ij} - \frac{d_i d_j}{2\mathcal{W}}) \delta(c_i, c_j). \tag{3.5}$$

where $d_i = \sum_j \mathcal{A}_{ij}$ is the degree of i and $2\mathcal{W} = \sum_j d_j$ is the total number of edges in the network. $\delta(c_i, c_j) = 1$ if $c_i = c_j$ and 0 otherwise and c_i (c_j) is the cluster or community to which vertex i (j) belongs.

This measure has high value when more edges in a graph fall between vertices of the same group than one would expect by chance. The main goal is to find good divisions of a graph into communities by optimizing $\mathcal{M}$ over possible divisions.

The matrix $\mathbf{C}$,

$$\mathbf{C}_{ij} = \mathcal{A}_{ij} - \frac{d_i d_j}{2\mathcal{W}} \tag{3.6}$$

in Eq. (3.5), refers to the *node centrality matrix*, and it plays a role in the maximization of the modularity equivalent to that played by the Laplacian in standard spectral clustering. Unlike the Laplacian, however, the eigenvalues of the centrality matrix are not necessarily all of one sign, and in practice, the matrix usually has both positive and negative eigenvalues.

Let μ denote the number of positive eigenvalues (and corresponding eigenvectors) of the centrality matrix $\mathbf{C}$, then the maximum number of possible communities is given by $(\mu + 1)$ [25]. The eigenspectrum of the centrality matrix $\mathbf{C}$ is closely linked to the community structure of the graph. Therefore, to reveal the community structure of the similarity graph, we proceed as follows: First, we retain the μ eigenvectors corresponding to the largest positive eigenvalues. Then, we iterate over $j = 1, \ldots, \mu$, spanning the whole range of possible communities. In each iteration, (i) we run a K-means algorithm on the retained eigenvectors looking for a partition into $c = j + 1$ communities and (ii) we compute the corresponding modularity $\mathcal{M}(j)$ according to Eq. 3.5. Finally, we chose the optimal partition as the one with the maximum modularity $\max(\mathcal{M})$.

3.5 Experimental Setup

To evaluate the effectiveness of the proposed algorithms in addressing the problem of community detection in complex networks, a comprehensive experimental setup was devised. This section outlines the methodology employed, including the dataset utilized, parameter settings, evaluation metrics, and computational infrastructure.

3.5.1 Network Selection

To validate our approach, we have examined four real networks drawn from the social science literature, which have become established benchmarks for evaluating community structure detection algorithms. These networks are as follows:

- The Karate Club Network [28] depicts the interpersonal relationships within a karate club at a US university in the 1970s. Comprising 34 members, the network reflects the club's division into two factions following an internal dispute.
- The Bottlenose Dolphins Network [20] is derived from observations of a community of 62 bottlenose dolphins. Interactions between dolphin pairs were established through statistically significant frequent associations. Bottlenose dolphins exhibit fission-fusion societies, allowing individuals to make decisions regarding group membership. The network's community structure reveals two major communities: one comprising 21 dolphins and the other further subdividing into three subcommunities encompassing the remaining dolphins.
- The Jazz Bands Network [12] encompasses 198 bands active between 1912 and 1942. Bands are linked if they share at least one musician. This network underscores the presence of two large communities, with the largest community further branching into two subcommunities.
- The American Football Games Network [11] represents games between Division IA colleges during the Fall 2000 regular season. In this network, nodes represent teams and edges represent games between teams. The network exhibits a distinct community structure, with nodes partitioned into conferences, each containing approximately 8 to 12 teams. Games are more frequent within the same conference than between teams from different conferences.

Moreover, we have tested the performance of our method on computer-generated graphs with a known community structure. We adopted the early class of standard benchmark graphs introduced by Lancichinetti et al. [18], based on the benchmark graphs of Girvan and Newman [11]. These graphs consist of $N = 128$ nodes split into four communities of 32 nodes each. Links between nodes belonging to the same community are drawn with probability p_{in}, while pairs belonging to different communities are joined with probability p_{out}. The value of p_{out} is chosen so that the average number of inter-community edges per node, denoted as z_{out}, can be controlled. Each node has, on average, z_{in} edges to nodes in the same community and z_{out} edges to nodes in other communities, maintaining a total average node degree $k = z_{\text{in}} + z_{\text{out}} = 16$.

We consider values of z_{out} ranging from 5, corresponding to a clear community structure ($z_{\text{out}} \ll z_{\text{in}}$), to 10, which describes a network with poorly defined structure ($z_{\text{out}} \gg z_{\text{in}}$). Since the "real" community structure is well known for these trial networks, we can validate our method by computing the fraction of correctly identified nodes.

3.5.2 *Dynamics of the Rössler Oscillators*

In this work, we study an ensemble of many interacting Rössler systems coupled through the y component:

$$\begin{aligned} \dot{x}_i(t) &= -\omega_i y_i - z_i, \\ \dot{y}_i(t) &= \omega_i x_i + a y_i + \tfrac{\sigma}{k_i} \textstyle\sum_j^N \mathcal{A}_{ij}(y_j - y_i), \\ \dot{z}_i(t) &= b + (x_i - c) z_i. \end{aligned} \tag{3.7}$$

All Rössler oscillators are nonidentical, and the parameter $\omega_i = \omega_0 + \Delta\omega_i$ selected randomly from a uniform distribution corresponds to the natural frequency of the individual oscillator. $\Delta\omega_i$ is the frequency mismatch between neighboring chaotic oscillators. We set $a = 0.2$, $b = 0.2$, $c = 5.7$, and $\omega_0 = 0.9$ so as to ensure the individual oscillator generates chaotic dynamics with phase coherent attractor.

As reported in Sect. 3.3, the oscillators are interconnected through a predefined network topology described by the $N \times N$ adjacency matrix $(\mathcal{A}_{ij})$. The strength of the coupling between the i-th and the j-th oscillator is given by σ if they are connected in the graph; otherwise it is zero. Finally, k_i is the degree of the ith oscillator within the network. The y-coupling would guarantee the stability of the synchronized state for a coupling force large enough [4].

Each oscillator is characterized by its natural characteristic frequency ω_i which distinguish each oscillator. Values of ω_i are randomly chosen from a uniform distribution with a frequency mismatch $\Delta\omega_i = 0.2$, so that the Rössler oscillators will evolve in a chaotic phase-coherent regime.

3.6 Numerical Results

In this section, we present numerical results obtained from the application of the adaptive frequency tuning algorithm followed by the modularity maximization algorithm to the real-world networks described in Sect. 3.5.1.

Numerical simulations were conducted, simulating the system over 100 trains of time series, each with a duration of 75 time units. We divided the simulation process into two phases. During the first 20 trains, every time unit, the coupling force is increased linearly from an initial value of $\sigma = 0$ up to a level below the threshold for inducing complete synchronization. Specifically, the attained coupling strengths were $\sigma_Z = 1.58$, $\sigma_D = 1.89$, $\sigma_J = 2.24$, and $\sigma_F = 1.03$ for the Zachary, Dolphins, Jazz, and Football networks, respectively. Throughout this phase, the coupled oscillators underwent synchronization based on the network topology and interaction strength, adjusting their dynamics to achieve coherence.

In the subsequent phase, the frequency tuning process, as outlined in Algorithm 1, was initiated. At each iteration, the oscillator frequencies were adjusted using the adaptation mechanism described above. During this phase, the modularity $\mathcal{M}$ was computed to quantify the effectiveness of partitioning the network into distinct communities, offering insights into the network's structural organization and dynamics.

Below, we report the main findings derived from our preceding study employing this methodology, as documented in [21]. We comprehensively present the outcomes encapsulating both average similarity and instantaneous modularity across the dynamic evolution of the network, thereby explicating the concept of dynamic frequency tuning.

Figure 3.1 shows the global temporal coherence of the whole network. Initially, in the absence of frequency adaptation, network coherence sharply increases due to the gradual increase of the coupling force. Notably, during this phase, oscillators within each final group or community tend to maintain synchronization despite their differing characteristic frequencies, underscoring the influential role of network topology in shaping synchronized groups. Subsequently, upon implementing the adaptation mechanism, a subtle improvement in the network's global coherence is observed, signifying the emergence of synchronized communities oscillating at a uniform average frequency.

The temporal evolution of modularity (see Fig. 3.2a) also follows a similar pattern, reflecting the impact of the frequency tuning process. We observe a pronounced rise in temporal modularity values $\mathcal{M}(t)$, which quickly reaches a plateau, indicating the effectiveness of the adaptation mechanism in fostering network coherence. As

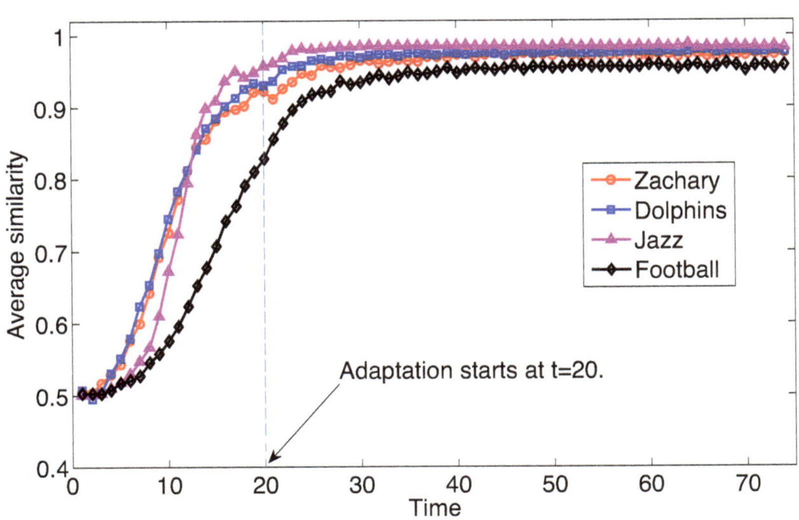

Fig. 3.1 The average similarity of the oscillator networks versus time performed over 100 runs of the algorithm. Adapted from reference [21]

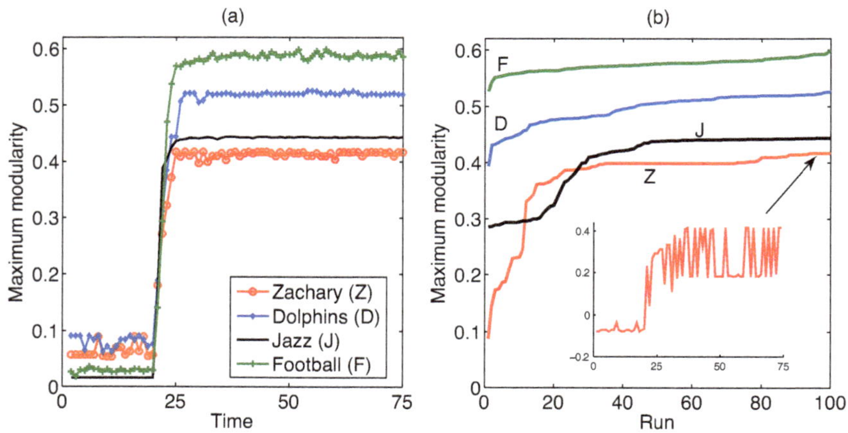

Fig. 3.2 (**a**) Time evolution of the maximum modularity during the network evolution. (**b**) Maximum values of modularity corresponding to each trial and reflecting the variability over the 100 different realizations. For Zachary and Dolphins, the maximum modularity corresponds to a division into four communities. For the Jazz network, the optimal division reveals three communities, while for the Football network, the optimal modularity is achieved for a partition into eight communities. In the inset of panel (**b**) temporal modularity of the Zachary network corresponding to one of the optimal runs. Adapted from reference [21]

the adaptation mechanism dynamically adjusts the frequencies of oscillators based on their interactions with neighbors, it fosters the formation of more coherent and distinct communities within the network. This process leads to a higher degree of intra-community connectivity and a corresponding decrease in inter-community connections, thereby enhancing the modular structure of the network.

The rapid increase in modularity values during the initial phase of frequency adaptation is characterized by the emergence of distinct communities of nodes with similar frequency dynamics, indicative of their functional specialization or shared behavioral patterns. As the frequency adaptation process continues, the rate of increase in modularity gradually slows down, eventually reaching a plateau. This plateau signifies the stabilization of the network's modular structure.

Additionally, upon examining the variance in maximum modularity values across successive iterations of the algorithm (refer to Fig. 3.2b), it becomes evident that attaining peak modularity, as delineated in Table 3.1, often necessitates only a small number of iterations.

Table 3.1 Comparison of modularity values for network division to the Girvan and Newman approach. Each entry displays the achieved modularity value along with the corresponding number of communities identified in parentheses

Method	Zachary	Dolphins	Jazz	Football
Girvan-Newman [11]	0.4090 (4)	0.4580 (4)	0.4379 (4)	0.5470 (6)
Moujahid et al. [21]	**0.4174** (4)	**0.5220** (4)	**0.4437** (4)	**0.5746** (6)

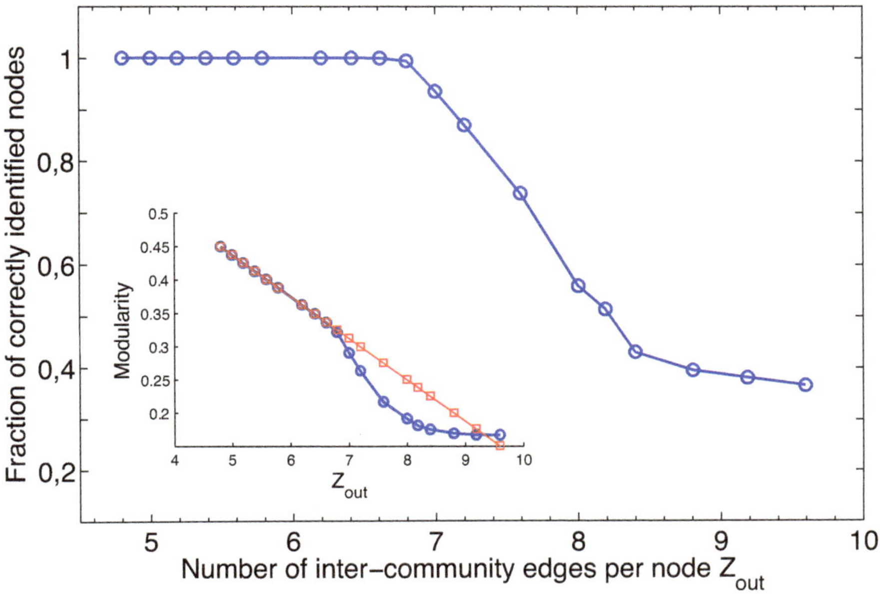

Fig. 3.3 Relationship between the fraction of correctly identified nodes and the average number of inter-community edges per node Z_{out}, averaged over 25 graph realizations. **(Inset)** Modularity $\mathcal{M}$ values achieved by our approach (circle markers) compared to those corresponding to the real partitions (square markers) versus Z_{out}. Adapted from [21]

Finally, we present experimental results conducted on trial networks, as described in Sect. 3.5. Figure 3.3 illustrates the values of the fraction of correctly identified nodes, averaged across 25 different realizations of the computer-generated networks, plotted against the average number of inter-community edges z_{out}. Additionally, we record the achieved modularity $\mathcal{M}$ values using our approach (see inset of Fig. 3.3), comparing them with those obtained using the synchronization-based dynamical clustering algorithm (OCR-HK) [5]. Notably, our approach demonstrates enhanced performance as soon as z_{out} surpasses six inter-modular edges per node, a threshold identified in Fig. 1 of Ref. [5]. While the sensitivity of the OCR-HK algorithm begins to decline beyond $z_{out} \geq 6$, our method maintains performance until $z_{out} > 7$. Moreover, in scenarios with clear community structures ($z_{out} < 6$), our method achieves complete performance, correctly predicting 100% of nodes, compared to approximately 90% by OCR-HK. For $Z_{out} > 7$, both algorithms achieve a similar fraction of correctly classified nodes. It is worth noting that at $Z_{out} = 8$, only a few algorithms can still identify over 80% of nodes correctly [8], albeit with high computational costs.

3.7 Conclusions

This chapter introduces a synchronization-based dynamic approach aimed at tackling the challenge of community detection in complex networks. The first contribution centers on adaptive frequency tuning for coupled oscillators, aiming to achieve synchronized states while preserving the diversity of the frequency vector. By dynamically adjusting oscillator frequencies based on network interactions, this approach promotes synchronization and clustering within the network structure, without converging to a homogeneous frequency vector. The algorithm iteratively updates oscillator frequencies, gradually increasing coupling strength to facilitate adjustment while leveraging similarity and dynamic connectivity matrices. This adaptive mechanism enables the algorithm to adapt to the network's topology and dynamics, ultimately enhancing its effectiveness in identifying community structures.

Similarly, the modularity maximization approach aims to extract an optimal partition from the network. By computing centrality matrices, retaining eigenvectors corresponding to the largest positive eigenvalues, and performing K-means clustering, the algorithm partitions the network into communities with high modularity. By comparing achieved modularity values with those obtained from real partitions, this algorithm provides insights into the quality of the identified communities. Through the combination of these two algorithms, our work offers a comprehensive framework for community detection in complex networks, providing robust solutions that adapt to diverse network structures and dynamics. These algorithms represent significant contributions to the field, offering scalable and effective methods for analyzing and understanding the complex organization of real-world networks.

Certainly, an avenue for further enhancement lies in fine-tuning the parameters of our approach to achieve even more robust and accurate results. This optimization process could involve adjusting key parameters such as the simulation time, the starting point of the adaptation process, and the frequency of adaptation steps. By systematically exploring the effects of these parameters on the network dynamics and community detection performance, we can identify optimal configurations that maximize the effectiveness of our methodology. Additionally, incorporating advanced optimization techniques or exploring alternative adaptation mechanisms could offer further improvements in capturing the evolving dynamics of complex networks. Thus, future research efforts could focus on refining the parameterization of our approach to unlock its full potential in network analysis and community detection tasks.

Acknowledgments This work is supported by grant PID2021-126701OB-I00 funded by MCIN/AEI/10.13039/501100011033 and by "ERDF A way of making Europe."

Competing Interests The authors have no conflicts of interest to declare that are relevant to the content of this chapter.

References

1. Int. J. Mod. Phys. C **16**, 515 (2005)
2. Almendral, J., Leyva, I., Li, D., Sendia-Nadal, I., Havlin, S., Boccaletti, S.: Dynamics of overlapping structures in modular networks. Phys. Rev. E **82**, 016115 (2010)
3. Arenas, A., Diaz-Guilera, A., Perez-Vicente, C.: Synchronization reveals topological scales in complex networks. Phys. Rev. Lett. **96**, 114102 (2006)
4. Boccaletti, S.: The synchronized dynamics of complex systems. In: *Monographs in Nonlinear Science and Complexity*, vol. 6 (2008)
5. Boccaletti, S., Ivanchenko, M., Latora, V., Pluchino, A., Rapisarda, A.: Detecting complex network modularity by dynamical clustering. Phys. Rev. E **75**, 045102(R) (2007)
6. Chen, J., Yuan, B.: Detecting functional modules in the yeast protein–protein interaction network. Bioinformatics **22**(18), 2283–2290 (2006)
7. Clauset, A., Newman, M., Moore, C.: Finding community structure in very large networks. Phys. Rev. E **70**, 066111 (2004)
8. Danon, L., Diaz-Guilera, A., Duch, J., Arenas, A.: Comparing community structure identification. J. Stat. Mech: Theory Exp. **2005**(09), P09008 (2005)
9. Flake, G., Lawrence, S., Giles, C., Coetzee, F.: Self-organization and identification of web communities. Computer **35**(3), 66–71 (2002)
10. Fortunato, S.: Community detection in networks. Phys. Rep. **486**, 75–174 (2010)
11. Girvan, M., Newman, M.: Community structure in social and biological networks. Proc. Natl. Acad. Sci. USA **99**, 7821 (2002)
12. Gleiser, P.M., Danon, L.: Community structure in Jazz. Adv. Complex Syst. **06**(04), 565–573 (2003). DOI 10.1142/s0219525903001067. http://dx.doi.org/10.1142/S0219525903001067
13. Guimera, R., Amaral, L.: Functional cartography of complex metabolic networks. Nature **433**(7028), 895–900 (2005)
14. Hegselmann, R., Krause, U.: Consensus and ragmentation of opinions with a focus on bounded confidence. Amer. Math. Monthly **126**(8), 700–716 (2019)
15. Hlaoui, A., Wang, S.: A direct approach to graph clustering. In: Proceedings of the IASTED International Conference on Neural Networks and Computational Intelligence, pp. 158–63 (2004)
16. Krause, A., Frank, K., Mason, D., Ulanowicz, R., Taylor, W.: Compartments revealed in food-web structure. Nature **426**(6964), 282–285 (2003)
17. Kuramoto, Y.: Chemical Oscillations, Waves and Turbulance. Springer, Berlin (1984)
18. Lancichinetti, A., Fortunato, S., Radicchi, F.: Benchmark graphs for testing community detection algorithms. Phys. Rev. E **78**(4), 046110 (2008)
19. Li, D., Leyva, I., Almendral, J., Sendia-Nadal, I., Buld, J., Havlin, S., et al.: Synchronization interfaces and overlapping communities in complex networks. Phys. Rev. Lett. **101**, 168701 (2008)
20. Lusseau, D., et al.: Behavioural ecology and sociobiology. Behav. Ecol. Sociobiol **54**, 396–405 (2003)
21. Moujahid, A., d'Anjou, A., Cases, B.: Community structure in real-world networks from a non-parametrical synchronization-based dynamical approach. Chaos, Solitons & Fractals **45**(9), 1171–1179 (2012). https://doi.org/10.1016/j.chaos.2012.06.007. https://www.sciencedirect.com/science/article/pii/S0960077912001312
22. Newman, M.: Networks: An Introduction. Oxford University, Oxford (2010)
23. Newman, M., Girvan, M.: Finding and evaluating community structure in networks. Phys. Rev. E **69**, 026113 (2004)
24. Newman, M.E.J.: Finding community structure in networks using the eigenvectors of matrices. Phys. Rev. E **74**, 036104 (2006)
25. Newman, M.E.J.: Modularity and community structure in networks. Proc. Natl. Acad. Sci. **103**(23), 8577–8582 (2006)

26. Pluchino, A., et al.: Modules identification by a dynamical clustering algorithm based on chaotic rössler oscillators. In: American Institute of Physics, Conference Proceedings, vol. 965, p. 323 (2007)
27. Rosvall, M., Bergstrom, C.: Maps of random walks on complex networks reveal community structure. Proc. Natl. Acad. Sci. USA **105**, 1118 (2008)
28. Zachary, W.: An information flow model for conflict and fission in small groups. J. Anthropol. Res. **33**, 452–473 (1977)

Chapter 4
Automatic Evolutionary Clustering for Human Activity Discovery

Daphne Teck Ching Lai and Parham Hadikhani

Abstract Clustering is regarded as a good approach to distinguish between different human activities from skeletal data in an unsupervised manner (also known as human activity discovery) because it does not require the laborious task of labeling a huge volume of data. In this chapter, we demonstrate a multi-objective evolutionary clustering methodology using particle swarm optimization, game theory, and Gaussian mutation techniques for performing such a task. The proposed methodology does not require any parameter setting nor prior knowledge of the number of clusters. It uses an automatic segmentation method based on kinetic energy to reduce redundant frame and identify keyframes. Features that characterize human motion are extracted from these keyframes and their dimensions are reduced using principal component analysis (PCA) before performing clustering on the reduced dataset. The proposed methodology was tested on popular benchmark datasets such as Cornell activity dataset (CAD-60), Kinect activity recognition dataset (KARD), Microsoft Research (MSR), Florence3D (F3D), and Nanyang Technological University (NTU-60) and compared with four automatic and four nonautomatic clustering algorithms, outperforming the other algorithms in most datasets. We demonstrate that the application of game theory enabled our clustering methodology to find the global best which is the optimal solution based on the multi-objective functions. We also showed that our methodology converges quickly due to the effects of game theory and Gaussian mutation.

Keywords Data clustering · Human activity discovery · Particle swarm optimisation · Multiobjective optimisation

---

D. T. Ching Lai (✉)
School of Digital Science, Universiti Brunei Darussalam, Gadong, Brunei Darussalam
e-mail: daphne.lai@ubd.edu.bn

P. Hadikhani
Department of Biomedical Informatics, University of Pittsburgh School of Medicine, Pittsburgh, PA, USA

UPMC Hillman Cancer Center, University of Pittsburgh School of Medicine, Pittsburgh, PA, USA
e-mail: parham.hadikhani@pitt.edu

F. Dornaika et al. (eds.), *Advances in Data Clustering*,
https://doi.org/10.1007/978-981-97-7679-5_4

4.1 Introduction

Clustering is the grouping of data points with similar characteristics together, where similarity is defined by a suitable metric such as the most commonly used Euclidean distance. Clustering allows us to identify patterns in the data without the laborious task of data labeling. It finds patterns that represent the characteristics of the groups in the data. Here, a set of C cluster centers $\mathbf{v} = \{v_1, \dots, v_C\}$ represents the clusters. Within a dataset, each cluster contains similar data points $Z = \{z_1, \dots, z_N\}$ with N number of data points and D dimensions, where each data point is a vector $\mathbf{z} = \{z_1, \dots, z_D\}$. The goal of clustering is to learn the partition matrix $\mathbf{U}$ of a dataset. The partition matrix shows that an object z_j belongs to the cluster C_i and is represented by a $C \times N$ matrix as $\mathbf{U} = [u_{ij}]$, where $i = 1, \dots, C$ and $j = 1, \dots, N$ such that $\sum_{i=1}^{C} u_{ij} = 1$. In hard clustering, such as K-means, u_{ij}=1 if $z_j \in C_i$ and 0, otherwise. These conventional techniques minimize a single objective function as follows:

$$J = \sum_{i=1}^{C} \sum_{j=1}^{N} u_{ij}^2 ||z_j - v_i||^2 \tag{4.1}$$

Minimizing a single objective function may not capture the different properties of hidden structures found in the dataset well [15]. Methods to simultaneously optimize multiple objective functions within an evolving population of solutions have been proposed [23]. In this manner, the clustering problem can be more accurately calculated, allowing for better-evolved solutions. Applications of multi-objective evolutionary algorithms (MOEA) in clustering have been shown to be effective in identifying meaningful clusters [15, 16, 23].

Such advanced clustering techniques are of growing interest in human activity recognition because they do not require labeled data to develop their model. Labeling data requires huge effort, especially when the volume of the data is large.

In this chapter, we present a multi-objective evolutionary clustering methodology for identifying human activities in an unsupervised way (multi-objective particle swarm optimization-based grouping of multi-objective task—MOPGMGT) [8], focusing on 3D skeleton data and not sensor-based data nor RGB videos for practical and privacy reasons. The methodology uses multiple objective particle swarm optimization (PSO) for clustering and uses Nash equilibrium (NE) to map potential solutions into a game theory (GT) space to improve the search for diverse clustering solutions. A validity index was applied to determine the optimal number of clusters. These processes are arranged in a pipeline to perform automatic evolutionary clustering. The idea of applying clustering to human activity discovery is due to its nature as an unsupervised learning technique. Like a baby discovering its new movement/activity, they does not know the name of the different activities but may repeat these activities again. Equally, this can be applied to a robotics problem such as human activity discovery to discover different activities without any human intervention. We also discuss other clustering techniques used to solve human activity discovery. This chapter demonstrates how data clustering is applied effectively in a methodology of human activity discovery, together with feature extraction techniques.

4.2 Human Activity Discovery Using Clustering

Human activity recognition (HAR) has emerged as a pivotal field with diverse applications spanning human-computer interaction, intelligent transportation systems [6, 42, 43], and monitoring applications [3]. The primary goal of HAR is to automatically identify and categorize human actions and activities in various environments. Traditional vision-based HAR systems commonly process visual data, but the inherent complexities of such data, including cluttered backgrounds, variations in brightness, and changes in perspective, pose challenges to system performance. To overcome these challenges, 3D skeleton data has gained prominence, offering a viable alternative that preserves privacy when RGB data capture is not feasible. Each frame, represented by the 3D coordinates of main body joints, provides a suitable representation for human actions [28]. This information can be obtained in real-time using low-cost depth sensors [13]. As depicted in Fig. 4.1, vision-based HAR systems typically involve several key steps. Vision sensors capture human activities, and skeletal information is extracted from video sequences. Meaningful features are then extracted to enhance activity discovery, and unsupervised clustering is employed to differentiate observed activities based on feature similarities. These discovered activity clusters serve as the basis for learning models for each activity, facilitating the recognition of future activities. While supervised learning of activity models has seen considerable progress, with reliance on human-labeled training data, we focus on the less-explored domain of activity discovery, referred to as human activity discovery (HAD) (Fig. 4.1 blocks (d) to (e)). This phase involves categorizing activities based on similarities without prior knowledge of activity labels, akin to a child learning from unlabeled data. The methodology presented in this study addresses the challenges associated with feature extraction and unsupervised clustering during this crucial stage of HAR. In the subsequent sections, we delve into the details of our methodology, which includes preprocessing, feature extraction, and automatic multi-objective clustering. These techniques collectively contribute to efficient HAD, even when the cluster number is unknown. This approach stands out as it operates on untreated, unsegmented input data, making it particularly relevant for real-world scenarios where activities are not predefined or annotated.

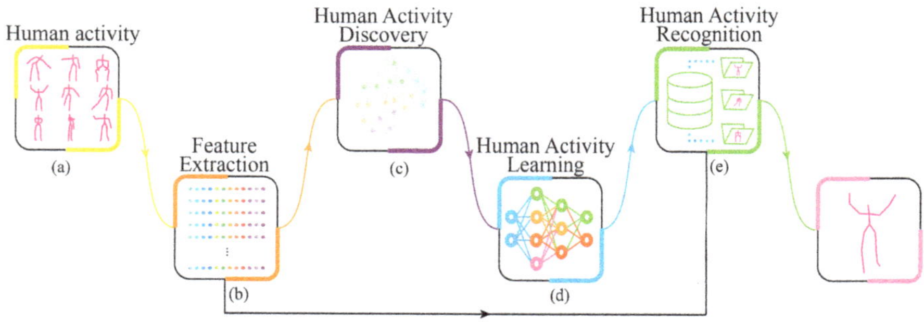

Fig. 4.1 Conceptual framework for human activity recognition system: (**a**) RGB-D sensor captures input frames and converts them to skeletal data. Then, (**b**) features are extracted from skeleton data, (**c**) activities are clustered based on the similarities and differences of features. Next, the system (**d**) learns the model for each activity based on clusters obtained in discovery step and finally (**e**) human activities can be recognized [10]

4.2.1 Preprocessing and Feature Extraction

In the initial stages of video analysis, not all recorded frames are relevant. Redundant information can impede activity discovery efficiency and increase computational requirements. To address this, we employ a kinetic energy calculation for each frame (f_i), based on joint movements, as defined by Eq. (4.2) from [9]. This approach involves calculating the movement of each joint (j) between frames i and i-1, with the sum of these movements representing the frame's kinetic energy. Keyframes, representing significant video content, are then selected based on the maximum and minimum local energy values [1].

$$E(f_i) = \sum_{j=1}^{J} E(f^j i) = 1/2 \sum j = 1^J (f^j i - f^j i - 1)^2 \tag{4.2}$$

To enhance the raw data representation and extract pertinent features for activity analysis, three categories of features—displacement, statistical, and orientation—are employed. Informative joints, including both left and right hands, feet, hips, shoulders, elbows, and knees, are selected to focus on relevant body movements.

Displacement features comprise spatial and temporal components. Spatial displacement involves computing the Euclidean distance between joints, while temporal displacement captures instantaneous changes between consecutive frames and neutral frames, resulting in a total of 72 features. Statistical features encompass mean and standard deviation differences of joint coordinates within a sequence, providing 72 features to distinguish between lower and upper torso activities.

Orientation features describe body posture and involve extracting rotation angles between selected bones. A rotation matrix is employed to calculate these angles relative to the x, y, and z axes, resulting in 21 orientation features. Additionally, angle features capture angles between specific bones, yielding four features.

Following feature extraction, principal component analysis (PCA) is applied to reduce data dimensionality while retaining 85 % of the variance. The sequence of frames is then partitioned into overlapping fixed-sized activity instances, each consisting of 15 frames. Overlapping instances enhance clustering performance by preserving information during transitions between activities [30]. Keyframe selection, feature extraction, and dataset preparation techniques, briefly outlined here, are detailed in [9].

4.2.2 Particle Swarm Optimization (PSO)

Particle swarm optimization (PSO) is a population-based algorithm used for optimization problems. Each individual particle in the population represents a potential solution. The position of each particle is updated iteratively using its velocity to search for the optimal solution [7]. The position and velocity of each particle are updated using the following equations:

$$x_i(t+1) = x_i(t) + v_i(t) \tag{4.3}$$

$$\begin{aligned} v_i(t+1) = w \times v_i(t) + c_1 \times rand_1 \times (pbest_i(t) - x_i(t)) \\ + c_2 \times rand_2 \times (gbest(t) - x_i(t)) \end{aligned} \tag{4.4}$$

where $x_i(t)$ is the position of particle i at time t, $v_i(t)$ is the velocity of particle i at time t, $pbest_i(t)$ is the local best position of particle i, $gbest(t)$ is the global best position of the population, w is the inertia weight, c_1 and c_2 are acceleration coefficients, and $rand_1$ and $rand_2$ are random values in [0,1]. The inertia weight w is adjusted over time to balance between exploration and exploitation:

$$w = \frac{w_{\max} + t \times (w_{\max} - w_{\min})}{t_{\max}} \tag{4.5}$$

where $w_{\max}$ and $w_{\min}$ are the maximum and minimum values of the inertia weight, respectively, and $t_{\max}$ is the maximum number of iterations. The acceleration coefficients c_1 and c_2 are also adjusted over time to control the effect of personal best and global best positions:

$$c_1(t+1) = (c_{1_{\max}} - c_{1_{\min}}) \times \frac{t}{t_{\max}} + c_{1_{\max}} \tag{4.6}$$

$$c_2(t+1) = (c_{2_{\max}} - c_{2_{\min}}) \times \frac{t}{t_{\max}} + c_{2_{\max}} \tag{4.7}$$

where $c_{1_{\max}}$, $c_{1_{\min}}$, $c_{2_{\max}}$, and $c_{2_{\min}}$ are the maximum and minimum values of the acceleration coefficients [2].

4.2.3 Automatic Multi-Objective Clustering Based on Game Theory

Hadikhani et al. [8] applied multi-objective clustering based on game theory on HAD. In multi-objective clustering, datasets are clustered based on multiple properties simultaneously, unlike in one-objective clustering where only one property is considered. This approach aims to optimize solutions across different criteria. The general form of a multi-objective problem is:

$$\textbf{Minimize } f_i(x) : i = 1, \ldots, N \tag{4.8}$$

where f_i is the ith objective function, N is the number of objective functions, and x is the decision vector.

One common objective function in multi-objective clustering is the sum of squared errors (SSEs), which aims to minimize the distance between data points and their respective cluster centroids:

$$SSE = \sum_{k=1}^{K} \sum_{\forall x \in c_k} \|x_i - \mu_k\|^2 \tag{4.9}$$

where x_i is a data point belonging to the cluster c_k and μ_k is the mean of the cluster c_k. Another objective function often used is the connectivity index (Conn-index), which measures the connectivity within clusters:

$$Conn = \frac{\sum_{i=1}^{k} \min \sum_{j=1}^{n} d(p_j^i, m_j)}{n(\min_{i,j=1,i \neq j}^{k} d(m_i, m_j))} \tag{4.10}$$

$$m_i = \min_t^n (\sum_{i=1}^{k} d(p_t^i, p_j^i)/n) \tag{4.11}$$

where n is the number of objects in cluster c_i, p_j^i is the jth object of cluster i, and m_i is the minimum distance of objects in cluster i to all other clusters' objects. In multi-objective clustering, several optimal solutions, called Pareto-optimal sets or non-dominated solutions, are obtained. These solutions are stored in a predefined repository, and when the repository is full, less important solutions are removed using the roulette wheel selection method. To maintain diversity in solutions, Gaussian mutation, as proposed in [9], is applied to non-dominated solutions:

$$v'_{non-dominated_i}(d) = v_{non-dominated_i}(d) \times G(0, h) \times (x_{\max}(d) - x_{\min}(d)) \tag{4.12}$$

$$x'_{non-dominated_i}(d) = x_{non-dominated_i}(d) + G(0, h) \times v'_{non-dominated_i}(d) \tag{4.13}$$

where $x'_{non-dominated_i}(d)$ and $v'_{non-dominated_i}(d)$ represent the position and velocity of ith non-dominated particle in the dth dimension. $x_{\max}$ and $x_{\min}$ are the maximum and minimum values, respectively, in the dth dimension. G is a Gaussian distribution with mean 0 and variance h. h is linearly decreased at each iteration according to:

$$h(t+1) = h(t) - (1/t_{\max}) \tag{4.14}$$

to balance exploration and exploitation throughout the iterations. To determine the global best solution, game theory (GT) is employed to make decisions regarding the Pareto-optimal sets by calculating the optimal strategy under specific circumstances to maximize outcomes. A game theory space comprises players on opposing sides, strategies (actions taken by players at different stages), and payoffs (consequences of strategies). To align the multi-objective problem with GT, each objective function is treated as a player, particles represent players' strategies, and the fitness value of each objective function serves as a payoff. Nash equilibrium (NE) is applied to identify the global solution from the Pareto-optimal set. Each player selects the best feasible strategy based on their interests, without collusion or assistance from others. To select the best global optimal solution from the Pareto-optimal set, the following equations from [18] are used to compute NE and find the global best solution:

$$NashE_j = \sum_{i=1}^{N} OBJ_{ji} \tag{4.15}$$

$$OBJ_{ji} = \frac{currentOBJ_{ji} - BestOBJ_i}{BestOBJ_i} \tag{4.16}$$

where N represents the size of the particle population. $NashE_j$ denotes the Nash equilibrium (NE) criterion of the jth individual and is intended to be minimized, $currentOBJ_{ji}$ stands for the ith objective of the jth particle and $BestOBJ_i$ indicates the best fitness value of the ith objective.

To determine the optimal number of clusters (activities) without prior knowledge, the Jump method [35] is applied. The proposed multi-objective clustering algorithm is initially executed for different values of k in the range from $k_{\min}$ to $k_{\max}$, where

k_{min} is set to 2 and k_{max} is determined based on $\sqrt{n}$, where n is the number of data points. Subsequently, Jump is computed for each value of k within the specified range using Eq. (4.17), and the best value of k with the minimum amount of Jump is selected.

$$Jump(k) = \frac{1}{m} \times \min E\left[\sum_{j=1}^{k} \sum_{x_i \in c_j} (x^j i - cj)^T \times \Gamma^{-1} \times (x^j i - cj)\right] \quad (4.17)$$

In this equation, x_i^j represents a data point with p dimensions in cluster c_j, and Γ is the within-cluster covariance matrix. The Jump method computes the optimal number of clusters by minimizing the expected within-cluster sum of squares. The routine of the proposed clustering algorithm is outlined in Algorithm 2.

4.2.4 Results and Discussion

The MOPGMGT method was validated on six datasets: Nanyang Technological University (NTU) RGB+D dataset (NTU-60) [33], Cornell activity dataset (CAD-60) [36], Kinect activity recognition dataset (KARD) [5], Microsoft Research (MSR) DailyActivity3D [37], UTKinect-Action3D (UTK) [38], and Florence3D (F3D) [32]. Each experiment was repeated 30 times, with a swarm size of 20 and 50 iterations per run. The maximum and minimum values for cognitive and social parameters ($c1_{max}, c2_{max}, and c1_{min}, c2_{min}$) were 2.5 and 0, respectively, and the maximum inertia weight (w_{max}) was 0.9. The stop criteria were based on the number of iterations. The performance of MOPGMGT was compared with four automatic clustering algorithms including PSO [24], hierarchical partitioning-based grouping of multi-objective tasks (HPGMK) [9], MOPSO (multi-objective PSO) [4], multi-objective PSO with Gaussian mutation (MOPGM), and four nonautomatic clustering algorithms (Kmeans, subspace clustering, elastic net subtractive clustering, and sparse subspace clustering—KM, SC, ENSC, and SSC, respectively [28]).

The clustering algorithms were evaluated based on accuracy [11], precision rate, F-score, and confusion matrix across 30 runs. The overall error (OE) [12] of estimating the number of clusters was also computed for each method.

Figure 4.2 indicates the minimum, maximum, and average accuracy of the proposed algorithm and other algorithms on all subjects of each dataset. The overall average accuracy of MOPGMGT was 72.43 % for CAD-60, 47.41 % for KARD, 36.78 % for MSR, 52.06 % for UTK, 56.43 % for F3D, and 35.43 % for NTU-60. MOPGMGT outperformed other algorithms in four datasets: CAD-60, MSR, F3D, and NTU-60 in terms of maximum and average accuracy. However, MOPGM performed slightly better than MOPGMGT in KARD and UTK datasets. It is noteworthy that, unlike KM, SC, ENSC, and SSC, MOPGMGT has no prior knowledge of the number of clusters but automatically estimates them and has been

Algorithm 2: Automatic multi-objective clustering based on game theory algorithm

```
Input:
    D={d1,d2,...,dn} //Set of data points
    Kmax //Maximum number of cluster
        //calculated by sqrt(size(D))
    Kmin ← 2
Output:
    The best clustering result
    Number of clusters
Steps:
for Kmin = 2 → Kmax do
    Randomly generate initial positions and velocities of particles
    Calculate the cost of each particle using Eq. (4.15) and (4.16)
    Set the initial personal best of each particle
    Repository ←non-dominated solutions
    Global best ←none
    iter ← 0, MaxIteration ← 50
    while iter ≺ MaxIteration do
        for each solution in Repository do
            Calculate the NashE_j for non-dominated solutions using Eq. (4.12)
        Global best ← find solution with the best NashE_j
        for each particle do
            Update particle velocity and position
            Calculate the new cost of each particle Eq. (4.15) and (4.16)
            Update the particle personal best
        if Size (repository size) ≻ numberofparticles/2 then
            Until size (repository size) ≻ numberofparticles/2, delete one Repository
            members based on "roulette wheel selection"
        for T times do
            Mutate solutions in repository according to Eq. (4.9) and (4.10)
            Compare the mutated version with the previous one and choose the
            non-dominated one
        iter ← iter+1
    Select the best solution using Nash equilibrium
Select the best value of k with the minimum value of Jump
```

able to outperform all the other approaches in most datasets. NE has led to the best decision of choosing the most optimal solution in MOPGMGT.

Another advantage of the proposed method over other methods is the use of Gaussian mutation to modify the solutions, allowing more areas of the search space to be covered and explored. This has resulted in better performance of the proposed method than the other methods, as well as preventing early convergence and generating more diverse solutions. Moreover, compared to HPGMK and PSO, which used a single objective function, MOPGMGT considered two aspects of good solutions in clustering, compactness, and connectivity, by using two different objec-

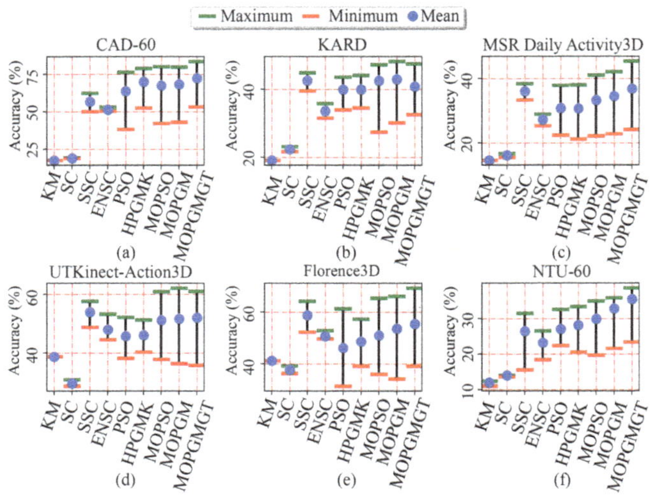

Fig. 4.2 The average accuracy for all subjects in (**a**) CAD-60, (**b**) KARD, (**c**) MSR (**d**) UTK, and (**e**) F3D [8]

Fig. 4.3 Illustration of global best selection from non-dominated solution by two methods of (**a**) game theory selection and (**b**) roulette wheel selection for KARD in one run [8]

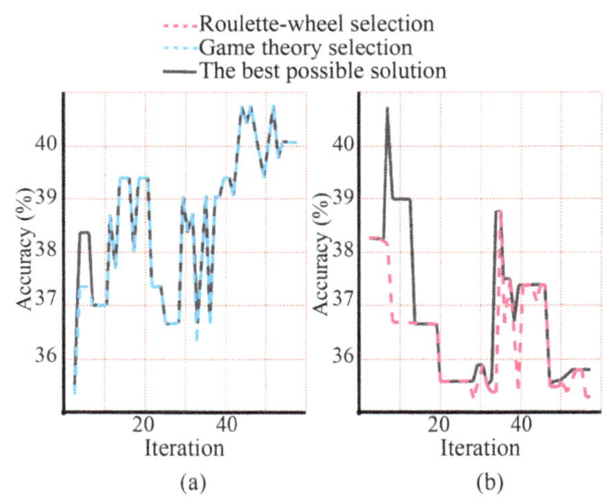

tive functions. It enabled MOPGMGT to capture diverse properties of activities and cluster them appropriately. KM and SC had the worst results due to being stuck in the local optimum. ENSC and SSC have had poor results compared to MOPGMGT. These methods lack an efficient strategy to balance exploration and exploitation.

Figure 4.3 illustrates the selection of the global best solution (continuous gray line) from various solutions on the Pareto front using two methods: game theory (blue dotted line) and roulette wheel selection (pink dotted line). This experiment, conducted with prior knowledge of the best solution (gray line) in each iteration, aims to demonstrate the impact of game theory on selecting the best solution. The results indicate that game theory (Fig. 4.3a) performs significantly better in selecting

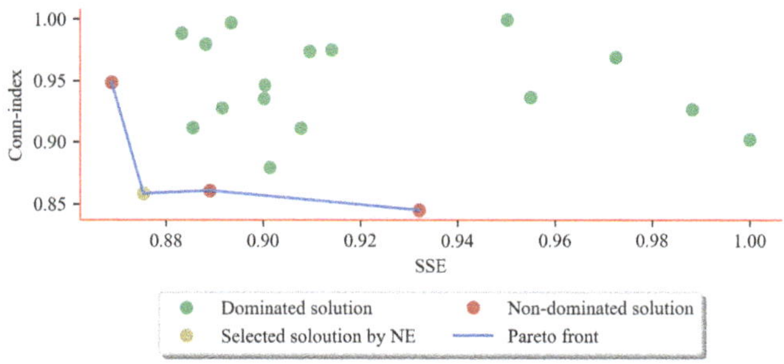

Fig. 4.4 Illustration of the final Pareto front obtained by MOPGMGT for subject 8 of F3D dataset [8]

the global best solution from the Pareto front compared to MOPSO, which utilizes an adaptive grid and roulette wheel selection [4].

Game theory demonstrates superior performance by identifying the best solution in nearly all iterations, as opposed to roulette wheel selection in Fig. 4.3b, which frequently fails to find the global best solution. The strength of game theory lies in its approach, which not only aims to find the global best solution but also considers the impact of selecting one solution on the other solutions. It chooses a solution that optimizes the overall state of all solutions. Furthermore, Fig. 4.3 highlights that game theory not only finds the global best solution but also improves the accuracy of the solution incrementally with each iteration. In contrast, the roulette wheel selection method struggles to find the global best solution due to its lack of a specific strategy and consideration for the consequences of selecting each solution.

Figure 4.4 illustrates the final Pareto front, showcasing the non-dominated solutions and the optimal solution chosen by MOPGMGT. The selected optimal solution (global best, displayed in a yellow dot) exhibits a superior position compared to dominated solutions (green dots), and the non-dominated solutions (red dots) are closer to the ideal point of both objectives than the other solutions. It is noteworthy that the solutions selected as non-dominated are significantly closer to the global optimum in both objectives than other dominated solutions. This is because when NE is utilized as a decision-maker and selects the global solution, it not only seeks the best optimum solution but also considers the impact of that selected solution on other solutions, thereby improving non-dominated solutions as well.

Table 4.1 presents the results of estimating the number of clusters for different methods using the Jump index. The best results are highlighted in bold, and in cases where the best results were achieved by several methods, they are italicized. The best results are those that are closer to the actual number of clusters, while the worst results are the opposite. As shown in the table, methods demonstrate varying performances in estimating the number of clusters across different datasets. However, the proposed algorithm achieves the lowest OE due to its superior ability to find the global best solution using NE compared to the MOPSO and MOPGM methods, as well as its utilization of multi-objective optimization for clustering

Table 4.1 Estimated number of clusters for different approaches using Jump index for five datasets

Dataset	#clusters	PSO	HPGMK	MOPSO	MOPGM	MOPGMGT
CAD-60	14	19	*16*	17	*16*	*16*
KARD	18	20	19	20	**18**	19
MSR	16	12	22	20	18	**17**
UTK	10	13	13	11	**10**	11
F3D	9	10	10	10	12	**9**
NTU-60	17	17	19	19	20	**22**
OE	–	4.561	4.491	4.457	4.413	**4.345**

Table 4.2 Impact of each component of FMOPG on the performance of HAD in KARD

Method	Accuracy (%)
PSO	42.1
MOPSO	45.4
MOPSO+GM	46.84
MOPSO+GM+GT	49.06

compared to PSO and HPGMK. In the NTU-60 dataset, the algorithms tend to consider a large number of activities as independent clusters due to the high number of activities and the similarity between many of them. Therefore, additional features need to be extracted to address this issue.

Table 4.2 presents an ablation study on the components of the MOPGMGT algorithm using the UTK dataset. When using single-objective clustering (PSO), only 42.1 % of the clusters are correctly identified. However, considering additional clustering factors such as compactness and separation and performing multi-objective clustering (MOPSO) improves clustering performance by approximately 3 %. Further enhancing the algorithm by applying Gaussian mutation to the non-dominated solutions (MOPSO+GM) focuses on optimizing solutions in a better area, resulting in an increased clustering accuracy of 46.84 %. However, the most significant improvement in clustering performance is observed when selecting the best solution, as shown in Fig. 4.3. By mapping the multi-objective space to the game theory space (MOPSO+GM+GT), the clustering accuracy reaches 52.06 %.

Figure 4.5 illustrates the impact of different components in our proposed method on algorithm convergence. The green line represents the convergence of a single objective optimized by PSO. However, since data qualities vary across different points, a single objective measure does not perform consistently well, leading to the worst convergence rate for PSO. MOPSO, depicted by the blue line, enhances the convergence rate compared to PSO by considering several objectives and diverse criteria. When Gaussian mutation is applied to MOPSO solutions (MOPGM), it increases the algorithm's ability to explore solutions, enhancing diversity and improving convergence. In essence, Gaussian mutation enhances the algorithm's exploitation ability. In the case of simple MOPSO, the roulette wheel selection method is used to determine the best solution. As shown in Fig. 4.3, roulette wheel

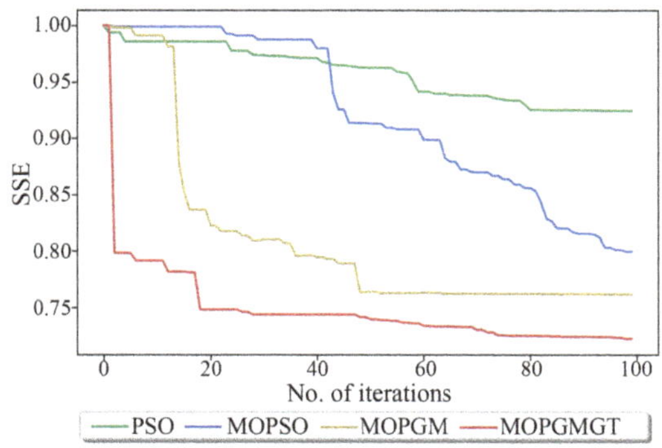

Fig. 4.5 Illustration of the proposed method's different components effect on the convergence. PSO algorithm is considered as the basic method. Then, the effect of multi-objectiveness (MOPSO) and increasing the diversity using Gaussian mutation (MOPGM), as well as the role of game theory (MOPGMGT) in algorithm convergence, are investigated [8]

Table 4.3 Comparisons of our HAD method with supervised and unsupervised HAR on the NTU RGB+D 60 in terms of accuracy

Types	Methods	Precision
Supervised	DT-LSTM 2012 [36]	25.5
	DeepRNN 2016 [33]	56.3
	HOPC 2014 [5]	50.1
	ST-LSTM 2016 [20]	55.7
	MS-G3D 2020 [22]	91.5
Unsupervised	Shuffle&Learn 2016 [25]	46.2
	Li et al. 2018 [17]	60.8
	LongT GAN 2018 [44]	39.1
	MS2L 2020 [19]	52.6
	CAE 2021 [31]	80.55
	P&C FW-AEC 2020 [34]	49.7
	TS Colorization2021 [40]	42.9
	PCRP 2021 [39]	54.9
	STAN 2022 [21]	48.3
	RGCA 2021 [41]	54.4
Discovery	SSC 2021 [28]	26.43
	HPGMK 2022 [9]	28.15
	MOPGMGT [8]	35.43

selection's performance is unstable, making it challenging to find the best solution consistently. However, when the multi-objective problem space is mapped to game theory (MOPGMGT) to determine the best solution, the proposed algorithm's convergence is improved by identifying the optimal solution more effectively.

Table 4.3 compares supervised and unsupervised methods for human activity recognition (HAR) with the proposed methods for human activity discovery (HAD). Supervised methods use training data and labels for all stages, including feature extraction and activity recognition. While unsupervised methods perform feature

Table 4.4 Comparison of the MOPGMGT with multi-objective clustering algorithms based on F-measure

Dataset	TSMPSO	MOPSO	NSGA II	MABC	MOVPS	MOPGMGT
Iris	0.927	0.909	0.890	0.860	0.947	**0.987**
Glass	0.558	0.544	0.555	0.502	0.569	**0.807**
Cancer	0.874	0.868	0.852	0.878	0.894	**0.867**
Wine	0.708	0.690	0.647	0.694	0.739	**0.886**
Vowel	0.632	0.611	0.609	0.602	0.692	**86.59**
Dermatology	0.366	0.358	0.359	0.362	0.372	**0.407**

extraction without labels, they still use training data for activity recognition. In contrast, HAD methods group activities based on feature similarities without using labels or training data for activity learning. The accuracy of supervised and unsupervised methods is generally higher than HAD methods due to their use of labeled data. However, the multi-objective particle swarm optimization-based grouping of multi-objective task (MOPGMGT) method stands out among HAD approaches. MOPGMGT employs two objectives to evaluate group compactness and separation, which leads to better results compared to the hierarchical partitioning-based grouping of multi-objective tasks (HPGMK) method. Additionally, MOPGMGT's simultaneous exploration and exploitation strategy improves its accuracy over methods like sequential sparse coding (SSC), which lacks a specific strategy for finding the optimal solution.

Table 4.4 compares the performance of MOPGMGT with other multi-objective clustering algorithms. This experiment focused solely on the clustering performance without involving feature extraction. The results demonstrate MOPGMGT's significant advantage over other methods across all datasets. The key differentiator is MOPGMGT's ability to balance objectives to reach the best possible solution. Unlike other algorithms that simply select the best solution from the Pareto front in each round without considering its impact on the rest of the solutions, MOPGMGT considers the effects of the selected solution on other solutions based on game theory. This balanced approach creates a win-win scenario between the SSE and Conn-index objectives, resulting in a solution that not only optimizes both objectives but also improves the overall optimization process. In essence, MOPGMGT's approach creates a mutually beneficial outcome for both objectives, ensuring that the selected solution is optimal for both SSE and Conn-index.

4.3 Other Clustering Techniques

Ong et al. [26] proposed the discovery of clusters (human activities) incrementally such that when one cluster is found using K-means, the samples belonging to the found clusters are omitted from the pool and the clustering iterates now with k decremented by 1, appying to CAD-60 dataset [36]. Their approach does not require prior knowledge of the number of clusters but the minimum number of samples in the cluster parameter, $MinPt$. They extracted all joints in all frames as features. For the HAR part, they applied Gaussian hidden Markov model to model the actions found by their incremental approach [27].

An encoder-decoder recurrent neural network, named predict & cluster (PC), capable of self-organizing the hidden structures into clusters of similar distinct movements was proposed [34]. All body keypoints sequences were preprocessed using view-invariant transformation and down-sampling to a maximum of 50 frames. The encoder is a multi-layered bidirectional gated recurrent unit (GRU) while the decoder is a uni-directional GRU. It was deemed a fully unsupervised HAR system though we do not know its HAD performance. The model was tested using K-nearest neighbor with $k = 1$ on NTU-60 [33] dataset achieving 39.6% accuracy.

Xu et al. [39] focused on unsupervised representation learning, developing a encoder-decoder based approach as an expectation-maximization (EM) task named prototypical contrast and reverse prediction (PCRP), where action prototypes are regarded as latent variables. The E-step uses K-means to cluster action encodings from the Uni-GRU encoder to determine prototype distribution. The M-step minimizes loss to optimize the encoder parameters. PCRP was demonstrated to outperform PC [34] on the NTU-60 and NTU-120 dataset.

A subspace clustering method which discriminates between activities using a covariance matrix and manages the temporal characteristics of the data using timestamp pruning was proposed for HAD [28]. The activity containing all joints within a number of timestamp is converted to a covariance matrix and flattened to represent one data point. Temporal pruning captures similarities over time with respect to kinematic information. An affinity graph matrix is obtained using temporal subspace clustering on the pruned skeleton data. However, the inputs to their system were pre-segmented and the method needed to specify the number of clusters.

4.4 Other Unsupervised (Non-clustering) HAR Techniques

There is a growing interest in developing sophisticated unsupervised feature or representation learning for HAR. This means addressing the human activity recognition problem without the human activity discovery part. By doing so, a classification layer which requires the use of labeled data is required, thus deeming the HAR system not fully unsupervised. The rationale here is that if meaningful input representation of human activities can be extracted in an unsupervised manner, it will work well to classify new activities. However, the classification part requires labeled data and thus, such human activity recognition system cannot be regarded as unsupervised. We discuss three such approaches for comparison with clustering-based approaches.

Zheng et al.[44] proposed a three subnetwork framework based on recurrent neural network and generative adversarial network containing the encoder, decoder, and discriminator networks, capable of handling whole input sequence with varying lengths and learning long-term global dynamics in an unsupervised manner. A classification layer at the end is added to classify the actions. This means it is not an end-to-end unsupervised human activity recognition system.

Paoletti et al. [29], who previously devised subspace clustering for human activity discovery, proposed an unsupervised feature extraction method for HAR based on a new convolutional autoencoder architecture that uses graph Laplacian regularization to model the skeletal characteristics across temporal dynamics of actions. To evaluate the features found, linear evaluation protocol and 1-nearest neighbor are used to train the classifiers.

A transformer-based contrastive learning method was developed for unsupervised representation learning of human activities to handle the frequency aspect using discrete cosine transform, self-attention module, multilayer aggregation perceptron, and spatiotemporal domain using spatiotemporal mining reconstruction module [14].

4.5 Conclusion

In this chapter, we discussed the application of clustering in human activity discovery. Human activity discovery is a prior step to distinguish between different activities without knowing the labels as opposed to human activity recognition which requires labeled data for different activities. An unsupervised human activity recognition system would be one using labels it generated during human activity discovery and not those generated by humans. We explained the methodology of applying multi-objective evolutionary clustering with game theory and Gaussian mutation to solve HAD problems and demonstrated its effectiveness. The model generated can be used with any classifiers to develop an unsupervised HAR system.

The practical applications of the methodology extend across several fields, with significant potential impact. In healthcare, our approach could be used for monitoring and analyzing patient activity patterns, aiding in early detection of health issues or providing personalized care plans. In sports science, it could help coaches and athletes analyze and improve performance by identifying optimal movement patterns. In smart environments, such as smart homes or offices, our methodology could enable context-aware systems that adapt to human activities, enhancing efficiency and user experience. Additionally, in security and surveillance, it could be used for abnormal behavior detection, enhancing public safety. Overall, the versatility and effectiveness of our methodology make it a valuable tool with broad applications in various industries and fields.

We also discussed other clustering methods that have been applied in this area as well as a few unsupervised techniques for HAD and HAR. We observed the trend that researchers are moving toward unsupervised feature or representation learning to solve HAR rather than on applying unsupervised clustering techniques to solve HAD first and then proceed to solve HAR fully unsupervised. By applying non-clustering approach in human activity recognition, labeled data is still required at the classification part at the human activity recognition stage (see Fig. 4.1), deeming them as supervised human activity recognition system. While the methodology has demonstrated effectiveness, there are still opportunities for further research. Future studies could explore integrating additional machine learning techniques like [11] or applying the methodology to different types of datasets. Our work contributes to the trend toward unsupervised feature or representation learning in human activity recognition, highlighting the potential for future advancements in this area.

References

1. Arzani, M.M., Fathy, M., Azirani, A.A., Adeli, E.: Switching structured prediction for simple and complex human activity recognition. IEEE Trans. Cybern. **51**(12), 5859–5870 (2020)
2. Cai, J., Wei, H., Yang, H., Zhao, X.: A novel clustering algorithm based on DPC and PSO. IEEE Access **8**, 88200–88214 (2020)
3. Chandrashekhar, H.V., et al.: Human Activity Representation, Analysis, and Recognition (2006)
4. Coello, C.A.C., Pulido, G.T., Lechuga, M.S.: Handling multiple objectives with particle swarm optimization. IEEE Trans. Evol. Comput. **8**(3), 256–279 (2004)
5. Gaglio, S., Re, G.L., Morana, M.: Human activity recognition process using 3-d posture data. IEEE Trans. Hum.-Mach. Syst. **45**(5), 586–597 (2014)
6. Hadikhani, P., Eslaminejad, M., Yari, M., Ashoor Mahani, E.: An energy-aware and load balanced distributed geographic routing algorithm for wireless sensor networks with dynamic hole. Wirel. Netw. **26**(1), 507–519 (2020)
7. Hadikhani, P., Hadikhani, P.: An adaptive hybrid algorithm for social networks to choose groups with independent members. Evol. Intell. **13**(4), 695–703 (2020)
8. Hadikhani, P., Lai, D.T.C., Ong, W.H.: Human activity discovery with automatic multi-objective particle swarm optimization clustering with gaussian mutation and game theory. IEEE Trans. Multimedia **26**, 420–435 (2023)
9. Hadikhani, P., Lai, D.T.C., Ong, W.H.: A novel skeleton-based human activity discovery using particle swarm optimization with gaussian mutation. IEEE Trans. Hum.-Mach. Syst. **53**(3), 538–548 (2023)
10. Hadikhani, P., Lai, D.T.C., Ong, W.H.: Flexible multi-objective particle swarm optimization clustering with game theory to address human activity discovery fully unsupervised. Image Vis. Comput. **145**, 104985 (2024). https://doi.org/10.1016/j.imavis.2024.104985. https://www.sciencedirect.com/science/article/pii/S0262885624000891
11. Hadikhani, P., Lai, D.T.C., Ong, W.H., Nadimi-Shahraki, M.H.: Improved data clustering using multi-trial vector-based differential evolution with gaussian crossover. In: Proceedings of the Genetic and Evolutionary Computation Conference Companion, pp. 487–490 (2022)
12. Hadikhani, P., Lai, D.T.C., Ong, W.H., Nadimi-Shahraki, M.H.: Automatic deep sparse multi-trial vector-based differential evolution clustering with manifold learning and incremental technique. Image Vis. Comput. **136**, 104712 (2023)
13. Han, J., Shao, L., Xu, D., Shotton, J.: Enhanced computer vision with microsoft kinect sensor: a review. IEEE Trans. Cybern. **43**(5), 1318–1334 (2013)
14. He, Z., Lv, J., Fang, S.: Representation modeling learning with multi-domain decoupling for unsupervised skeleton-based action recognition. Neurocomputing, **582**, 127495 (2024)

15. Hruschka, E.R., Campello, R.J., Freitas, A.A., et al.: A survey of evolutionary algorithms for clustering. IEEE Trans. Syst. Man Cybern. Part C Appl. Rev. **39**(2), 133–155 (2009)
16. Lai, D.T.C., Sato, Y.: Hybrid multiobjective evolutionary algorithms for unsupervised qpso, bbpso and fuzzy clustering. In: 2021 IEEE Congress on Evolutionary Computation (CEC), pp. 696–703. IEEE, New York (2021)
17. Li, J., Wong, Y., Zhao, Q., Kankanhalli, M.S.: Unsupervised learning of view-invariant action representations. Adv. Neural Inf. Proces. Syst. **31**, 1254–1264 (2018)
18. Li, X., Gao, L., Li, W.: Application of game theory based hybrid algorithm for multi-objective integrated process planning and scheduling. Expert Syst. Appl. **39**(1), 288–297 (2012)
19. Lin, L., Song, S., Yang, W., Liu, J.: Ms2l: Multi-task self-supervised learning for skeleton based action recognition. In: Proceedings of the 28th ACM International Conference on Multimedia, pp. 2490–2498 (2020)
20. Liu, J., Shahroudy, A., Xu, D., Wang, G.: Spatio-temporal lstm with trust gates for 3d human action recognition. In: European Conference on Computer Vision, pp. 816–833. Springer, Berlin (2016)
21. Liu, M., Bao, Y., Liang, Y., Meng, F.: Spatial-temporal asynchronous normalization for unsupervised 3d action representation learning. IEEE Signal Process Lett. **29**, 632–636 (2022)
22. Liu, Z., Zhang, H., Chen, Z., Wang, Z., Ouyang, W.: Disentangling and unifying graph convolutions for skeleton-based action recognition. In: Proceedings of the IEEE/CVF Conference on Computer Vision and Pattern Recognition, pp. 143–152 (2020)
23. Maulik, U., Bandyopadhyay, S., Mukhopadhyay, A.: Multiobjective Genetic Algorithms for Clustering: Applications in Data Mining and Bioinformatics. Springer Science & Business Media, Berlin (2011)
24. Van der Merwe, D., Engelbrecht, A.P.: Data clustering using particle swarm optimization. In: The 2003 Congress on Evolutionary Computation, 2003 (CEC'03), vol. 1, pp. 215–220. IEEE, New York (2003)
25. Misra, I., Zitnick, C.L., Hebert, M.: Shuffle and learn: unsupervised learning using temporal order verification. In: European Conference on Computer Vision, pp. 527–544. Springer, Berlin (2016)
26. Ong, W.H., Palafox, L., Koseki, T.: An incremental approach of clustering for human activity discovery. IEEJ Trans. Electron. Inf. Syst. **134**(11), 1724–1730 (2014)
27. Ong, W.H., Palafox, L., Koseki, T.: Autonomous learning and recognition of human action based on an incremental approach of clustering. IEEJ Trans. Electron. Inf. Syst. **135**(9), 1136–1141 (2015)
28. Paoletti, G., Cavazza, J., Beyan, C., Del Bue, A.: Subspace clustering for action recognition with covariance representations and temporal pruning. In: 2020 25th International Conference on Pattern Recognition (ICPR), pp. 6035–6042. IEEE, New York (2021)
29. Paoletti, G., Cavazza, J., Beyan, C., Del Bue, A.: Unsupervised human action recognition with skeletal graph laplacian and self-supervised viewpoints invariance. arXiv preprint arXiv:2204.10312 (2022)
30. Presti, L.L., La Cascia, M.: 3D skeleton-based human action classification: a survey. Pattern Recogn. **53**, 130–147 (2016)
31. Rao, H., Xu, S., Hu, X., Cheng, J., Hu, B.: Augmented skeleton based contrastive action learning with momentum lstm for unsupervised action recognition. Inf. Sci. **569**, 90–109 (2021)
32. Seidenari, L., Varano, V., Berretti, S., Bimbo, A., Pala, P.: Recognizing actions from depth cameras as weakly aligned multi-part bag-of-poses. In: Proceedings of the IEEE Conference on Computer Vision and Pattern Recognition Workshops, pp. 479–485 (2013)
33. Shahroudy, A., Liu, J., Ng, T.T., Wang, G.: Ntu rgb+ d: A large scale dataset for 3d human activity analysis. In: Proceedings of the IEEE Conference on Computer Vision and Pattern Recognition, pp. 1010–1019 (2016)
34. Su, K., Liu, X., Shlizerman, E.: Predict & cluster: unsupervised skeleton based action recognition. In: Proceedings of the IEEE/CVF Conference on Computer Vision and Pattern Recognition, pp. 9631–9640 (2020)

35. Sugar, C.A., James, G.M.: Finding the number of clusters in a dataset: an information-theoretic approach. J. Am. Stat. Assoc. **98**(463), 750–763 (2003)
36. Sung, J., Ponce, C., Selman, B., Saxena, A.: Unstructured human activity detection from rgbd images. In: 2012 IEEE International Conference on Robotics and Automation, pp. 842–849. IEEE, New York (2012)
37. Wang, J., Liu, Z., Wu, Y., Yuan, J.: Mining actionlet ensemble for action recognition with depth cameras. In: 2012 IEEE Conference on Computer Vision and Pattern Recognition, pp. 1290–1297. IEEE, New York (2012)
38. Xia, L., Chen, C.C., Aggarwal, J.K.: View invariant human action recognition using histograms of 3d joints. In: 2012 IEEE Computer Society Conference on Computer Vision and Pattern Recognition Workshops, pp. 20–27. IEEE, New York (2012)
39. Xu, S., Rao, H., Hu, X., Cheng, J., Hu, B.: Prototypical contrast and reverse prediction: unsupervised skeleton based action recognition. IEEE Trans. Multimedia **25**, 624–634 (2021)
40. Yang, S., Liu, J., Lu, S., Er, M.H., Kot, A.C.: Skeleton cloud colorization for unsupervised 3d action representation learning. In: Proceedings of the IEEE/CVF International Conference on Computer Vision, pp. 13423–13433 (2021)
41. Yao, H., Zhao, S.J., Xie, C., Ye, K., Liang, S.: Recurrent graph convolutional autoencoder for unsupervised skeleton-based action recognition. In: 2021 IEEE International Conference on Multimedia and Expo (ICME), pp. 1–6. IEEE, New York (2021)
42. Yari, M., Hadikhani, P., Asgharzadeh, Z.: Energy-efficient topology to enhance the wireless sensor network lifetime using connectivity control. Journal of Telecommunications and the Digital Economy **8**(3), 68–84 (2020)
43. Yari, M., Hadikhani, P., Yaghoubi, M., Nowrozy, R., Asgharzadeh, Z.: An energy efficient routing algorithm for wireless sensor networks using mobile sensors. arXiv preprint arXiv:2103.04119 (2021)
44. Zheng, N., Wen, J., Liu, R., Long, L., Dai, J., Gong, Z.: Unsupervised representation learning with long-term dynamics for skeleton based action recognition. In: Proceedings of the AAAI Conference on Artificial Intelligence, vol. 32 (2018)

Chapter 5
Identification of Correlated Factors for Absenteeism of Employees Using Clustering Techniques

Divyajyoti Panda, Debjani Panda, and Satya Ranjan Dash

Abstract Absenteeism has become one major concern for companies in this digitization era. To sustain in the competitive market and ensure timely delivery of products and services, as per the timeline, an uninterrupted performance of workforce is essential. This chapter focuses on finding out the crucial factors affecting the presence of the employee, and clustering has been used for the classification of absenteeism. The K-means algorithm has been used for determining clusters and critical correlated features have been mapped with a heatmap using a publicly available data set from University of California, Irvine (UCI). Factors related to lifestyle like obesity, higher BMI, employees with higher workload, distance of employee from workplace, etc. have been identified to be the most important factors affecting the presence of the employee at the workplace.

Keywords K-means · Classification · Clustering · Similarity functions · Heatmap

5.1 Introduction

In the era of digitization, companies have cutthroat competition in the market, and missing to deliver the target and deliverables on time may have a huge impact on the company's profitability and reputation in the market. The profitability and productivity of the company entirely depend on its workforce, i.e. its employees.

D. Panda
University of Southern California, Los Angeles, CA, USA
e-mail: divyajyo@usc.edu

D. Panda
Indian Oil Corporation Limited, Bhubaneswar, India
e-mail: pandad@indianoil.in

S. R. Dash (✉)
KIIT Deemed-to-be-University, Bhubaneswar, India
e-mail: sdashfca@kiit.ac.in

F. Dornaika et al. (eds.), *Advances in Data Clustering*,
https://doi.org/10.1007/978-981-97-7679-5_5

It is vital for every company to maintain a healthy workforce, who are mentally and physically in good health. Maintaining a healthy workforce goes a long way in ensuring there is minimum absenteeism in the company and that products and deliverables are offered as per timelines. Almost every company is critically affected by absenteeism of its employees, and critical projects remain unexecuted due to absenteeism of the employees. Employee absenteeism may occur due to illness, retirement, retrenchment, or any other factors. Absenteeism equally affects the employees and the organization. Whenever an employee superannuates or leaves the organization, the entire money and time that were involved in training and developing the employee are wasted, thus affecting the productivity and efficiency of the company. Quite a lot of projects, which were in the planning phase, may not get executed due to the absence of qualified manpower that possesses the requisite skills. Hence, there is a dire need to closely study the absenteeism of employees and the underlying reasons or causes, so that delays can be avoided, and projects can get smoothly executed. It is also essential to carefully choose the employees for important and critical projects, so that they are completed on time and do not tarnish the image of the company. Thus, it is crucial for every company to identify the crucial factors affecting the presence of an employee. Every employer needs to be aware of the causal factors of employee absenteeism so that they are in a position to take preventive measures to avoid such occurrences. This chapter focuses on the identification of crucial factors that have a high correlation with employee absenteeism. Clustering techniques are useful in studying large data sets for having hidden patterns. The critical factors have been identified by the K-means algorithm, and the correlated features have been plotted using heatmap. The study uses a data set of a courier company in Brazil and is available in the public domain. The main cause of absenteeism has been found out to be medical consultation by the employees and employee health conditions. A sedentary lifestyle, overweight, and age have been found to be the critical factors affecting the presence of the employee.

5.2 Definition of Clustering

Clustering is considered to be one of the most important concepts in unsupervised learning that is related to partitions within a data structure that can be further explored for learning. According to one of the definitions [1]: (i) Similarity should be more within instances of the same cluster. (ii) Differences should be more when clusters belong to different instances. (iii) The method of measurement should be clear and should be practical enough to implement and convey the desired meaning. Also, if we look at some other definitions, we come across the clustering process [2] wherein it is defined as possessing the steps like: (i) Extracting and selecting important features from the data set (ii) Designing the algorithm for problem resolution, as per the given set of characteristics of the problem (iii) Evaluating the result (iv) Explaining or standardizing results Depending upon the criteria, mentioned above, the similarity and differences(distances) are determined within

Table 5.1 Similarity and dissimilarity functions

Similarity/dissimilarity function	Formulae	Description
Jaccard similarity	$\frac{\|A \cap B\|}{\|A \cup B\|}$	It is the measure of similarity between two sets, where—X—represents the no. of elements and J Distance = 1 - J Similarity
Hamming similarity	It is the minimum number of substitutions required to change one data point into another data point	Similarity is greater when number is smaller
Mixed data type	$S_{ij} = \frac{\sum_{l=1}^{d} \eta_{ijl} S_{ijl}}{\sum_{l=1}^{d} \eta_{ijl}}$	Features are mapped into (0, 1). Attributes are transformed to dichotomous attributes
Euclidean distance	$\sqrt{\sum_{k=1}^{d} (x_{2k} - x_{1k})^2}$	Weighted euclidean distance obtained from deviation
Cosine distance	$\frac{\boldsymbol{x}_1^T \boldsymbol{x}_2}{\\|\boldsymbol{x}_1\\|_2 \\|\boldsymbol{x}_2\\|_2}$	Remains unchanged on rotation of data
Pearson correlation distance	$\frac{(\boldsymbol{x}_1 - \overline{\boldsymbol{x}_1})^T (\boldsymbol{x}_2 - \overline{\boldsymbol{x}_2})}{\\|\boldsymbol{x}_1 - \overline{\boldsymbol{x}_1}\\|_2 \\|\boldsymbol{x}_2 - \overline{\boldsymbol{x}_2}\\|_2}$	It calculates distances in linear correlation
Minkowski distance	$\left(\sum_{k=1}^{d} \|x_{2k} - x_{1k}\|^p\right)^{\frac{1}{p}}$	For n=1, distance is city-block distance, for n=2 distance is euclidean, for n= infinity, distance is chebyshev
Mahalanobis distance	$\sqrt{(\boldsymbol{x}_1 - \boldsymbol{x}_2)^T \boldsymbol{S}^{-1} (\boldsymbol{x}_1 - \boldsymbol{x}_2)}$	Covariance matrix S with higher complexity

instances of the clusters using various mathematical functions. Some of them are listed in Table 5.1.

5.3 Clustering Techniques

5.3.1 Distribution-Based Clustering

This is dependent on the fact that data is divided into clusters, where each cluster has a mean and a variance. It is a type of hierarchical clustering where the data points are distributed like Gaussian and normal distribution. This type of distribution helps in identifying outliers in large data sets, making large data sets handling easier, and aids in identifying non-linear relationships. The concept of this type of clustering states that the data points belonging to a distribution belong to the same clusters, even if there are several distributions in the data set. The algorithms considered for distribution-based clustering are distribution-based clustering of large spatial databases (DBCLASD) [3] and gaussian mixture model (GMM) [4].

5.3.2 Density-Based Clustering

In this type of clustering, data points are identified as clusters based on the density of the data set, which means the clusters are identified based on their concentration. Wherever they are close together, the data points are identified clusters, and they have no clear boundaries. Density-based spatial clustering of applications with noise (DBSCAN) is an example of density-based clustering algorithm. It executes by allocating each point with a minimum number of neighbours before making it a part of a cluster.

5.3.3 Partition-Based Clustering

In partition-based clustering, the data points are divided into K different clusters, and the algorithm continues to reallocate the data points to the nearest clusters which continues till a local optimal partition is obtained. The method involves allocating the data points uniquely to one of the clusters. Each of the clusters has one cluster representative or a medoid. There is a high degree of similarity within the data points of the cluster. The centroid is normally called the medoid and is calculated by the mean of the points present within the cluster. This is one of the popular methods of clustering, and examples include K-means algorithm, partition around medoids (PAM) (K-medoid), and CLARA algorithms.

5.3.4 Hierarchical-Based Clustering

Hierarchical as the name suggests is a method to generate a group of clusters that are nested within clusters. The structure replicates that of a tree and can be visualized as a dendrogram—in which the tree-like structure denotes the separations and joints by visibly clear demarcations. Hierarchical procedures can be either divisive or agglomerative. The agglomerative method begins with each element denoting a single cluster and joins them in sequences to form larger clusters, whereas divisive methods begin with the entire group and proceed further by splitting the group into successively minor clusters.

5.3.5 Fuzzy-Based Clustering

Fuzzy clustering is a method where the values are not discrete. It is useful for continuous values, where each data point has some degree of belongingness to each cluster. Hence, at any given point in time, the data point is measured by its degree of belongingness to the cluster. It has its values in decimals and not in 0s and 1s. The more the value of belongingness, the more the similarity of the data point with the cluster.

5.3.6 *Categorization of Model-Based Clustering*

Model-based clustering method is used for the optimization of data, and these are mostly assumption-oriented mathematical models. These models are purely based on assumptions, and they try to find the optimum allocation of data points to their respective clusters.

5.3.7 *Grid-Based Clustering*

Grid-based clustering is performed for data points in a value space, and not the data points. This technique uses multi-resolution grid data structure. The method initially starts by covering the problem space domain with a uniform grid mesh. Clustering is performed on the grid instead of data points, by collecting statistical attributes of all the data objects that are located in each individual mesh cell. These algorithms are faster in processing than other algorithms, since they scan the data set only once to compute the statistical values for the grids, and the performance of clustering depends only on the size of the grids, which in most cases is lesser than the data objects. Some of the grid-based clustering algorithms that are used include STING [5], WaveCluster [6], and CLIQUE [7]. All these methods employ a uniform grid mesh to cover the whole problem. For problems with highly irregular data distributions, the resolution of the grid mesh must be fine enough to obtain a good clustering quality. A finer mesh can result in the mesh size close to or even exceeding the size of the original mesh or grid. The prime advantage of grid clustering is its faster processing/computation time for very large data sets.

5.4 Related Works

A work by Karaboga et al. [8] describes the ABC, i.e. the artificial bee colony algorithm being utilized as an optimization function which aids in simulating the enthusiastic honey finding behaviour of a swarm of honeybees that is interpreted for classification and clustering of data points within a pool of available data. The proposed work deals with three categories of artificial colonies such as employed bees, scouts, and onlookers. The employed bee is considered to be associated with a definite source of food and offers the neighbourhood, on the basis of its memory. The other set of bees, i.e. the onlookers, help in obtaining the information regarding the source of food, from the bees employed in the hive, and try to identify a food source that is nearer to the nectar. The third category of bees, i.e. the scouts are accountable for finding new sources of nectar. Also, another basic information of this algorithm says that the number of food sources is equal to the number of employed bees in this hive. The chapter proposed the clustering algorithm, where each object is associated with K-clusters, by decreasing the sum of the distance between these objects. The ABC algorithm has been equated with other exploratory algorithms like particle swarm optimization (PSO) and genetic algorithm, considering performance as their metrics, and various data sets like glass identification, data sets of thyroid, and wine have been studied from UCI repository, for studying the performance of the model. A ratio of 72:25 has been followed for train and test data, and the proposed ABC Algorithm performs better than other available models and gives only 13.13 'I&' errors as compared to an error of 15.99 'I&' in PSO model. The proposed model gives higher performance and better-quality clusters compared to its competitors.

Another work by Senthilnath et al., [9] proposed the firefly algorithm (FA) as one of the superior nature-inspired optimization programs for clustering, which simulates the pattern of fireflies. The firefly algorithm executes in three steps wherein the following activities are distributed. During step1, there is a random distribution of agents in the search space. In the subsequent steps, the objects were distributed into classes, wherein the cluster centre is identified. The data considered for training and testing the unsupervised algorithm have been taken from 13 data sets available in the UCI repository. The data is distributed in clusters using train and test method. The performance is calculated by applying the percentage of classification error, i.e. ((CEP). The metrics used help in finding out the best algorithm which has found the optimum cluster centres. The FA performance has been found to be better than the other two optimization algorithms like PSO and ABC algorithms. It also focuses on the fact that the performance of clusters is dependent on the volume of train data set and impacts the generation of clusters.

The authors Kim et al., [10] stated algorithms like density-based clustering (DBSCAN) and OPTICS for identifying clusters. The DBSCAN algorithm has been useful in identifying highly dense regions of data points that are identified as clusters. The clusters are identified on the basis of density parameters within an arbitrary shape. Usually, parallelization is not possible in the MapReduce framework, but the authors have proposed a technique, named DBCURE-MR which is helpful in identifying various clusters parallelly by spreading out every core point. As compared to conventional density-based methods, where each cluster is identified separately, the proposed method tries to find clusters parallelly. Different data sets like CLOVER, WINDOW, and BUTTERFLY have been experimented with, to assess the performance of the recently developed algorithm. The methods DBCURE and DBCURE-MR help in identifying cluster centres effectively and also can scale up using MapReduce framework.

Another piece of work by Chen et al. [11] mentions about spectral parallel procedures that have been used in the field of computer vision and for retrieval of information. The work brings out the comparison of two categories of approximates, consisting of similarity matrix sparsing and Nyström approximation for detecting parallel spectral clusters within a distributed nature of matrix. The term spectral signifies the manipulation of eigenvectors, eigenvalues, singular vectors, and singular values. The suggested method is more successful in locating clusters, as compared to the traditional algorithms, like K-means. The algorithm calculates similarity matrix information for grouping data points into k-clusters. While constructing the matrix of sparse similarity, the author calculated the distance between nearest neighbors among all data points, modified the matrix symmetrically, and then computed the similarity. The three steps have been implemented by using MapReduce, a Google parallel computing framework. To reduce memory usage, the sparsification approach has been used to generate the sparse matrix. The Nystrom approximation function has been used to store similarity matrices. The work has been carried out on three different data sets which were Corel (images), RCV1 (documents), and Picasa-Web from Google. The spectral clustering method generates clusters more effectively than the K-means method.

In another work by Shim et al., [12] customer relationship management (CRM) was developed with sequential patterns with association rules for a shopping mall (online). Even sequential patterns and association rules were used to design a CRM, and during the year 2002, dot-com bubble burst was established. It consisted of several mini online shopping centres. Due to good customer relationship management, many small shopping centres became established with lower costs and possessed all online market characteristics. The authors in their work have tried to study the transactions of various customers of an online shopping mall. They were successful in proposing sequential patterns and the rules of association within them. The authors [6] have tried to classify the very important customers (VIP) on the basis of their frequency of purchase, recency of purchase, and monetary value of their purchase. On the basis of these three factors, the model classifies the customers as VIP and non-VIP customers. The study has been carried out with several data mining methodologies like logistic regression, artificial neural network (ANN), bagging decision trees, and decision trees. The work successfully identifies rules and patterns for VIP customers from the transactional data available and also defines the sequential rules [13–25].

A work by Miguéis et al., [14] tried to focus on the segmentation of lifestyle of customers by using data mining methods. The market flourishes with good customer relationships and is essential for any company to have loyalty base of its customers. In the world of tough competition, it becomes essential for a company to have good business relations with its customers. In the work, the authors have tried to extract statistics from a large transactional database and were successful in proposing a market segmentation method for retail customers, based on their living style and in turn their purchasing history.

The work by Sastry et al., [26] signifies the usage of clustering methods, for discovering the variation in product sales and for comparing the sales data for a certain time period. The clustering method is effective in categorizing items naturally under one umbrella that seems to have similar properties. The work has been carried out on steel products' annual sales data to analyse the volume of sales and its corresponding value with respect to additional dependent variables like customers, products, and quantities sold. It has been observed that there is cyclic demand which depends on factors like the profile of customers, offered price, discounts, taxes, etc. Clustering methodologies like K-means 'I&' EM (expectation-maximization) have been used for the study for exploring hidden and interesting patterns that can improve sales revenue.

Another work by Niknam et al., [27] describes the K-means clustering and its effectiveness in identifying and creating K-clusters and also its effectiveness in finding the localized optimal solution. The authors have projected a hybrid clustering algorithm using fuzzy adaptive particle swarm optimization, ant colony optimization method, and K-means method and have named it FAPSO-ACO-K method, which is a simple and effective method for creating clusters. It identifies global optimal solutions and helps in determining the centres of clusters without many errors.

Another piece of work by authors Nanda et al., [28] proposed a new concept of representing data of stock markets in unlike clusters, as a new methodology of data mining. Depending upon the investment amount, the clustering methodology was designed. The authors implemented K-means, fuzzy C-means, and self-organizing maps for clustering data of huge databases of Bombay Stock Exchange. The clustering methodology classifies stocks from various data sets, and the authors pointed out K-means as the efficient method in classifying the data.

While describing the fuzzy clustering, the authors Kaymak et al., [29] emphasized the use of such clustering to divide the data set into several groups, in order to describe each cluster within the data. In fuzzy clustering, the data point has some belongingness to each cluster, instead of perfectly having its place in one cluster. Fuzzy clustering has been useful in various spheres like finance and marketing. However, the problems of size and bulk and shape of clusters, proper allocation of data points, etc. still remain as issues to be addressed in fuzzy clustering.

A piece of work by Fallahpour et al., [30] has experimented 79 stock data sets of Iran by using the clustering approach to classify the stocks into number of clusters. They experimented with three different types of clustering algorithms, namely K-medoids, K-means, and X-means algorithms, and their performance was assessed by the application of intra-class inertia that distinguishes between the clusters by the density of the clustering method. The intra-class inertia factor establishes the superiority of K-means algorithm over than other two algorithms, and clusters were extracted effectively using indices like Davis-Bouldin and silhouette.

A work by author Ernest B. Akyeampong et al. [31] focuses on the duration or period of absenteeism of its employees. The work identifies the time period for which an employee is absent from their work. The authors identify illness as the most responsible factor and also try to bring out the seasons responsible for such illness. They figure out winter season as the most vulnerable season for its employees. The work lays focus on the winter season, where employees fall ill more frequently than the summer season. Their work also lays focus on the fact that during the winter season, infectious and communicable diseases like common cold and influenza have a higher rate of infecting employees than in summer. The employees are subjected to these diseases and remain absent from work. The authors also explain that the occurrences of employee absenteeism during summer seasons are due to their availing vacations, and not infectious diseases. If the annual average absence of employees is considered, it was observed that 30 percent more part-leaves were reported during the winter season, as compared to 20 percent less part-leaves availed by the employees during summers.

Another work by Morten Nordberg and Knut Røedhave [32] focused on three factors viz. business cycles, health insurance of employees, and their absenteeism. The authors in their work attempted to measure the outcome and impact of the economic environment on the absenteeism of the employee. To study the impacts without any influence of other factors, the authors tried to isolate the business cycle impacts during the stages that could affect employee absenteeism. In their work, they tried to condition the state of the business cycle, during the phase when an employee was about to proceed in sick leave.

The authors acquired the following facts:

(i) Enhancing the business cycle has a low impact on the resumption rate of employees who are absent and, on the other hand, has higher reversion rates for employees who have resumed their work already. (ii) They also brought out the fact that temporary absenteeism is like an investment in health, where the employee is taking care of self and is avoiding long absence in the future. (iii) The recommencement rate to work increases when an employee has exhausted all their sickness benefits and is also noticed that this recommencement is normally short-lived.

Likewise, another work of author Hackett J.D. et. Al. [33] stated absenteeism as 'the temporary interruption of work for not less than one entire working day initiative of the employee, when his/her presence is expected by the employer'. Similarly, the Encyclopaedia of Social Sciences states 'Absenteeism' as the time lost in industrial establishment by the absence of employees due to avoidable or unavoidable reasons. The time lost by the strikes or by lateness amounting "to an hour or two is not usually included".

In another work by Parvez et al. [34], as per author Dakely C.A. 'Absenteeism is the ratio of the number of production man-days or shifts lost to the total number of productions scheduled to work'. As per the definition, by Labour Bureau (1962), absenteeism is defined as loss of total shifts as a percentage of the total number of shifts planned for executing the work.

A study by Wolter H.J. Hassink & 'Pierre Koning [35] showed a statistically substantial difference in the patterns that were generated due to the absence of the employees, amongst the group of workers, and these workers had diverse eligibility statuses and were depending on their attendance annals. It was also determined by the fact whether the employee/worker had won a lottery or not. It was noticed that absenteeism was more among employees who had already won and were not further considered to take part in firm-wide absence that was associated with the lottery.

A piece of work by Ruchi Sinha [36] brought out the fact that a mere 4 percent of the total employees remain absent from their work, and that is primarily due to their individual reasons. These employees reported their high work satisfaction in their workplaces.

In another piece of research work by Kammoun et al. [37], various responsible factors were brought out that critically affected employees' presence in the workplace. Some of the crucial factors were identified to be health problems, stress, loneliness at workplace, non-cooperation from colleagues, etc. These factors showed a strong inter-dependency with employee's absenteeism.

Another work by author K.A. Hari Kumar et. Al. [38] brought out the fact of work resumption on exhaustion of sick leaves by employees. They studied the reasons for work resumption by employees and noticed that work resumption on exhaustion of sick leave or benefits was short-lived and again the employees tend to go on leave due to medical reasons.

The review of studies examines the psychometric properties of absence measures, as well as the relationship between absenteeism and personal attitudinal, and organizational variables factors. Studies that examine the link between absenteeism and turnover are analysed by the research's unit of analysis.

After reviewing the methodologies and identifying absenteeism as one of the major concerns for an establishment, we felt the need for classifying the data set using K-means clustering, identifying the correlated features, and plotting them on a heatmap.

5.4.1 Data Set Details

The data set for the study has been obtained from UCI repository [39]. The data pertains to employees of a courier company in Brazil. It has the patient ID and the reason for absence recorded in the @nd column named ICD. This column has values ranging from 1 to 28 depending upon the category and the reason for absence. The last column is the target variable, i.e. the hours of absenteeism. The data set has 740 rows and 21 attributes, and the 21st attribute is the target variable. A detailed description of attributes is given in Table 5.2:

Table 5.2 Absenteeism data set

Attributes	Description
ID	Employee ID
Reason for absence (ICD)	1: A kind of infectious/parasitic diseases
	2: Neoplasms
	3: Diseases of the blood and corresponding disorders affecting immunity
	4: Endocrine, nutritional, and metabolic diseases
	5: Mental and behavioural disorders
	6: Nervous system diseases
	7: Eye and adnexa disease
	8: Ear and mastoid disease
	9: Circulatory system disease

(continued)

Table 5.2 (continued)

Attributes	Description
	10: Respiratory disease
	11: Digestive system disease
	12: Skin and subcutaneous tissue disease
	13I: Musculoskeletal and connective tissue disease
	14: Genitourinary system disease
	15: Pregnancy, childbirth, and puerperium
	16: Perinatal disease
	17: Congenital malformations, deformations, and chromosomal abnormalities
	18: Symptoms, signs, and abnormalities that are not listed above
	19: Injury, poisoning, or any other external causes
	20: Morbidity
	21: Factors influencing health status and contact with health services
	And seven categories without (CID) patient follow-up (22), medical consultation (23), blood donation (24), laboratory examination (25), unjustified absence (26), physiotherapy (27), dental consultation (28)
Absence month	The month in which absence of the employee is recorded
Day	Day of week in which employee remains absent
Spells(seasons)	Four seasons are studied for recording the absenteeism of employee
	(Summer: 1, Autumn: 2, Winter: 3, Spring: 4)
Transportation expense	The expense incurred by the employee for commuting to workplace
Distance (km)—residence to work	The distance from work to home
Service period	Years of service rendered by the employee
Age	Age of the employee
Average workload/day	Workload in hrs on an average for each employee
Hit target	HIT target
Disciplinary failure	Whether employee is facing any disciplinary action
Education	Education level of employee
	(High school: 1, Graduate: 2, Postgraduate: 3, Master and doctor: 4)
Children	No. of children of the employee
Social drinker	Whether employee is a social drinker (Yes=1, No=0)
Social smoker	Whether employee is a social smoker (Yes=1, No=0)
Pet	If employee possesses any pets?
Weight	Weight of the employee
Height	Height of the employee
Body mass index	BMI of employee
Absenteeism time in hours	Total hours of absenteeism

5.5 Methods and Methodology

For classification of the absenteeism of employees, K-means algorithm has been used, and classification has been done by clustering the data points and finding out the most correlated attributes. These correlated attributes are also plotted using heatmap and the causal factors are studied.

5.5.1 K-Means Algorithm

This algorithm effectively quantifies vectors and is mostly used in signal processing. This method divides the (n) number of observations by the (k) number of clusters, where each partition is separate from each other with no overlapping. In this method, each observation is allocated to the cluster, where the mean distance is minimum, i.e. the allocation is made to the nearest cluster. The nearest mean is also known as the cluster centre or centroid and acts as a prototype of the cluster. This type of clustering is helpful in minimizing cluster variances within the cluster by taking the squared Euclidean distances instead of normal Euclidean distances. The algorithm can be described as follows: The observations are recorded as m1(1), ..., mk(1). K-means algorithm proceeds by alternating between two phases:[7] 1. Assignment phase: Each observation is assigned to a cluster with the nearest mean, which is calculated by taking least squared Euclidean distance.
where each xp is assigned to exactly one S(t), even if it could be assigned to two or more of them. 2. Update phase: The mean is recalculated for observations and then assigned to each cluster.

The algorithm converges when the assignments do not change further. This algorithm does not always give the optimum results.

5.6 Result Analysis and Conclusion

When we try to plot the heatmap (Fig. 5.1), with columns correlated to absenteeism, we see some strong correlation among some attributes. The columns have been studied in detail to find the most important factors that directly impact the absenteeism of an employee.

The attributes that have a strong correlation with absenteeism have been identified as workload, years of experience, BMI, education of the employees, employees with no children, with no pets, and the autumn season. It is very much evident from the heatmap that the employees who have very high workload tend to exhaust themselves and are on frequent leave. It is also observed that the employees who have worked for the company for long years, somewhere around 7–17 years, are availing more leaves than the less experienced ones. Other factors that were seen

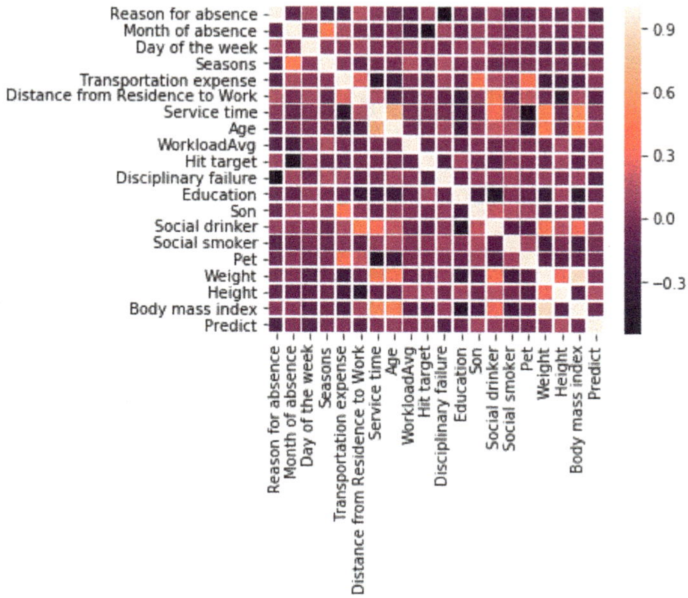

Fig. 5.1 Heatmap of correlated features affecting absenteeism

to affect absenteeism include the employee's health, i.e. higher BMI and obese employees remain absent more than the lower BMI employees. Since the data belong to a courier company and most of the employees are educated, it is seen that educated employees take more leaves and remain absent. Also, the employees who do not have any children or any pets tend to remain absent more than their counterparts. While studying these factors, it is also noticed that during the autumn season, absenteeism is higher as compared to other seasons. After identifying the critical factors, the company needs to take preventive measures as how to minimize the absenteeism of its employees, by eradicating the causal factors. The identified factors require the direct attention of the employer and thus aid the company in making decisions, to improve their productivity.

References

1. Jain, A.K., Dubes, R.C.: Algorithms for Clustering Data. Prentice-Hall, Inc., New York (1988)
2. Xu, R., Wunsch, D.: Survey of clustering algorithms. IEEE Trans. Neural Netw. **16**(3), 645–678 (2005)
3. Xu, X., Ester, M., Kriegel, H., Sander, J.: A distribution-based clustering algorithm for mining in large spatial databases. In: Proceedings of the Fourteenth International Conference on Data Engineering, pp 324–331 (1998)
4. Rasmussen, C.: The infinite Gaussian mixture model. Adv. Neural Inf. Process. Syst. **12**, 554–560 (1999)

5. Wang, W., Yang, J., Muntz, R.R.: STING: a statistical information grid approach to spatial data mining. In: The 23rd International Conference on Very Large Data Bases (VLDB'97), pp: 186–195 (1997)
6. Sheikholeslami, G., Chatterjee, S., Zhang, A.: WaveCluster: a multi-resolution clustering approach for very large spatial databases. In: The 24th International Conference on Very Large Data Bases (VLDB'1998), pp: 428–439 (1998)
7. Agrawal, R., Gehrke, J., Gunopulos, D., Raghavan, P.: Automatic Subspace Clustering of High Dimensional Data for Data Mining Applications. In International Conference Management of Data (SIGMOD'98), pp: 94–105 (1998)
8. Karaboga, D., Ozturk, C.: A novel clustering approach: artificial Bee Colony (ABC) algorithm. Applied Soft. Comput. **11**(1), 652–657 (2011)
9. Senthilnath, J., Omkar, S.N., Mani, V.: Clustering using firefly algorithm: performance study. Swarm Evol. Comput. **1**(3), 164–171 (2011)
10. Kim, Y., Shim, K., Kim, M.S., Lee, S.: DBCURE-MR: an efficient density-based clustering algorithm for large data using MapReduce. Inf. Syst. **42**, 15–35 (2014)
11. Chen, W.Y.,, Song, Y., Bai, H., Lin, C.J., Chang, E.Y.: Parallel spectral clustering in distributed systems. IEEE Trans. Pattern Anal. Mach. Intell. **33**(3), 568–586 (2011)
12. Shim, B., Choi, K., Suh, Y.: CRM strategies for a small-sized online shopping mall based on association rules and sequential patterns. Expert Syst. Appl. **39**(9), 7736–7742 (2012)
13. Stetco, A., Zeng, X., Keane, J.: Fuzzy cluster analysis of financial time series and their volatility assessment.In: IEEE International Conference in Systems, Man, and Cybernetics (SMC), pp. 91–96 (2013)
14. Miguéis, V.L., Camanho, A.S., e Cunha, J.F.: Customer data mining for lifestyle segmentation. Expert Syst. Appl. **39**(10), 9359–9366 (2012)
15. Pete, C., Julian, C., Randy, K., Thomas, K., Thomas, R., Colin, S., Rüdiger, W.: CRISP-DM. NCR,SPSS, DaimlerChrysler. USA, pp. 1–76 (2000)
16. Khan, M.N.A., Ullah, S.: A log aggregation forensic analysis framework for cloud computing environments. Computer Fraud Security **2017**(7), 11–16 (2017)
17. Khan, M.N.A., Wakeman, I.: Machine learning for post-event timeline reconstruction. In: First Conference on Advances in Computer Security and Forensics Liverpool, UK, pp. 112–121 (2006)
18. Rahman, S., Khan, M.N.A.: Review of live forensic analysis techniques. Int. J. Hybrid Inform. Technol. **8**(2), 379–88 (2015)
19. Khan, M.N.A., Chatwin, C.R., Young, R.C.: Extracting evidence from filesystem activity using Bayesian networks. Int. J. Forensic Comput. Sci. **1**, 50–63 (2007)
20. Khan, M.N.A.: Performance analysis of Bayesian networks and neural networks in classification of file system activities. Comput. Secur. **31**(4), 391–401 (2012)
21. Khan, M.N.A., Chatwin, C.R., Young, R.C.: A framework for post-event timeline reconstruction using neural networks. Digit. Investig. **4**(3–4), 146–157 (2007)
22. Bashir, M.S., Khan, M.N.A.: Triage in live digital forensic analysis. Int. J. Forensic Comput. Sci. **1**, 35–44 (2013)
23. Rafique, M., Khan, M.N.A.: Exploring static and live digital forensics: methods, practices and tools. Int. J. Sci. Eng. Res. **4**(10), 1048–1056 (2013)
24. Shehzad, R., Khan, M.N.A.: Integrating knowledge management with business intelligence processes for enhanced organizational learning. Int. J. Softw. Eng. Appl. **7**(2), 83–91 (2013)
25. Khalid, M., ul Haq, S., Khan, M.N.A.: An assessment of extreme programming based requirement engineering process. Int. J. Mod. Educ. Comput. Sci. **5**(2), 41 (2013)
26. Sastry, S.H., Babu, P., Prasada, M.S.: Analysis '&' Prediction of Sales Data in SA P-ERP System using Clustering Algorithms. arXiv preprint arXiv:1312.2678 (2013)
27. Niknam, T., Amiri, B.: An efficient hybrid approach based on PSO, ACO and k-means for cluster analysis. Appl. Soft Comput. **10**(1), 183–197 (2010)
28. Nanda, S.R., Mahanty, B., Tiwari, M.K.: Clustering Indian stock market data for portfolio management. Expert Syst. Appl. **37**(12), 8793–8798 (2010)

29. Kaymak, U., Setnes, M.: Extended fuzzy clustering algorithms. In: ERIM Report Series Reference No. ERS-2001-51-LIS (2000)
30. Fallahpour, S., Zadeh, M.H., Lakvan, E.N.: Use of clustering approach for Portfolio Management. Int. SAMANM J. Financ Acc. **2**(1), 115–136 (2014)
31. Akyeampong, E.B.: Trends and seasonality in absenteeism. Perspectives on Labour and Income **19**(3), 13 (2007)
32. Nordberg, M., Røed, K.: Absenteeism, health insurance, and business cycles. In: HERO WP, vol. 17 (2003)
33. Hackett, R.D.: New directions in the study of employee absenteeism: a research example (1986)
34. Parvez, A., Bhasin, J., Rasool, G.: Combined scale for measurement of job outcomes: psychometric properties and validation. J. General Manag. Res. **2017**(1), 36 (2017)
35. Wolters, J.C., Permentier, H.P., Meinema, A.C., Jansenss, G.E., Ciapaite, J., Heinemann, M., Veenhoff, L.M., Bakker, B.M., Bischoff, R.P.H.: Targeted proteomics as a tool to study biological pathways and processes. MCB2014. In: Joining Forces in Pharmaceutical Analysis and Medicinal Chemistry (2014)
36. Sinha, R., Rizvi, T.H., Chakraborti, S., Ballal, C.K., Kumar, A.: Primary melanoma of the spinal cord: a case report. J. Clin. Diagn. Res. JCDR **7**(6), 1148 (2013)
37. Kammoun, N., Dhifaoui, B.: Working conditions and employee absenteeism: a study on a sample of tunisian Agro-Food workers. Int. J. Sci. Res. (IJSR) **10**(2), 778–787 (2019)
38. Alexander, B.K., Moreira, C., Kumar, H.S.: Resisting (resistance) stories: a tri-autoethnographic exploration of father narratives across shades of difference. Qual. Inq. **18**(2), 121–133 (2012)
39. Martiniano, A., Ricardo, F.: Absenteeism at work. UCI Machine Learning Repository (2018). https://doi.org/10.24432/C5X882

Chapter 6
Multi-view Data Clustering Through Consensus Graph and Data Representation Learning

Fadi Dornaika and Sally El Hajjar

Abstract In the domain of multi-view clustering, existing methodologies can be categorized into subspace multi-view clustering algorithms, multi-view kernel approaches, matrix factorization approaches, and spectral clustering algorithms. However, a common limitation among these approaches is their dependence on combining predefined individual similarity matrices from multiple views. This susceptibility to noisy original similarity matrices, along with the integration of various spectral projection matrices, often affects their overall performance. To address these limitations, we introduce a novel approach named multi-view clustering with consensus graph learning and spectral representation (MCGLSR). In contrast to the traditional practice of directly integrating similarity matrices from different views, which may introduce noise, our proposed method simultaneously generates similarity graphs for each view and their shared similarity matrix (graph matrix) through a unified global objective function. This unified objective function ensures that the similarity matrices from different views are compelled to be sufficiently similar, effectively mitigating the impact of noise and promoting a more coherent unified data structure. Moreover, our approach facilitates the recovery of the common spectral projection and soft cluster assignments based on the shared graph structure. Crucially, MCGLSR operates on a kernelized representation of the views' features, producing individual graphs, a common graph, a common spectral representation, and cluster assignments directly. This eliminates the need for an external clustering algorithm in the final stage. To validate the efficacy of our technique, we conduct experiments on several real-world datasets, demonstrating its robust performance and addressing the identified shortcomings in existing approaches.

F. Dornaika (✉)
University of the Basque Country, San Sebastian, Spain

IKERBASQUE, Basque Foundation for Science, Bilbao, Spain
e-mail: fadi.dornaika@ehu.eus

S. El Hajjar
University of the Basque Country, San Sebastian, Spain

F. Dornaika et al. (eds.), *Advances in Data Clustering*,
https://doi.org/10.1007/978-981-97-7679-5_6

Keywords Multi-view clustering · Consensus graph learning · Consensus spectral representation matrix · Cluster assignment matrix

6.1 Introduction

In the contemporary landscape of machine learning, multi-view clustering techniques have garnered significant attention. Multi-view clustering approaches prove more effective as they leverage additional information from different views to comprehensively describe and categorize data. This stands in contrast to single-view clustering algorithms that rely on a singular feature to represent the data, a limitation often unsuitable in diverse scenarios [19, 29]. Among multi-view clustering methods, spectral clustering (SC) is widely utilized [16, 20, 27, 28]. Characterized by a well-defined mathematical framework, SC techniques are favored for their simplicity. Spectral clustering involves creating a graph that represents data points, allowing the data to be processed by a clustering algorithm using an appropriate spectral projection matrix. However, SC techniques are highly reliant on initialization and are sensitive to outliers due to their post-processing nature. Kernel techniques [22, 26] represent another category of clustering algorithms that map data into a space for easier separation. Despite their efficacy, these methods face challenges in dealing with large datasets due to their cubic computational cost. On the other hand, matrix factorization techniques [21, 23] are computationally efficient and often used for dimensionality reduction. Yet, many of these approaches directly merge similarity matrices from all views, potentially compromising efficiency with noisy graphs. In contrast, our study introduces a novel method called multi-view clustering using unified graph learning and spectral representation (MCGLSR). This method involves joint learning of individual graphs, a consensus graph matrix, a cluster-friendly consensus spectral representation, and a nonnegative cluster assignment matrix. The use of these consensus matrices addresses the challenge of merging different views with noisy data. An iterative optimization technique is employed to solve the objective function of the proposed approach. Extensive testing on various real-world datasets of different types and sizes demonstrates the superiority of our method compared to other state-of-the-art techniques.

To tackle the issue of noisy views represented by different similarity matrices, MCGLSR integrates a step for learning the consensus graph and spectral representation. By incorporating information from multiple views, the method mitigates the effects of noise in individual similarity matrices, enhancing clustering performance. This is achieved by enforcing similarity among individual matrices, resembling a unified similarity matrix "$\mathbf{S}^*$," which helps reduce the impact of noisy or unreliable information in individual matrices. Consequently, our proposed method can identify and filter out noisy information, leading to more robust clustering results. Moreover, simultaneous computation of multiple matrices ensures consistency among them. In other words, the computed matrices share common properties and represent the same underlying structure, resulting in improved clustering performance. The concurrent computation strategy employed by our method proves more efficient than computing each matrix separately, particularly when matrices are related and share common properties. Thus, the computational cost of multi-view clustering methods can be reduced, enabling more efficient processing of larger datasets.

The chapter makes the following contributions:

1. The proposed method jointly computes the graph of each view, the consensus graph, the corresponding spectral projection matrices, the consensus spectral representation matrix, and the nonnegative embedding matrix, all starting from either raw data or the corresponding kernel matrices. The nonnegative embedding matrix represents the cluster index matrix.
2. It directly provides cluster assignments from the soft cluster assignments without the need for any external clustering algorithm or post-processing.
3. It inherits the advantages and strengths of consensus multi-view learning techniques, matrix factorization strategies, and graph-based learning techniques.
4. It compares the proposed method with current multi-view techniques on various real multi-view datasets of different types and sizes.

The structure of the chapter is as follows: Sect. 6.2 introduces the main notations used and provides an overview of related work. In Sect. 6.3, the proposed approach is described. The optimization procedure for the proposed method is outlined in Sect. 6.4. Section 6.5 presents meaningful experimental results using real multi-view datasets to evaluate the efficiency of the proposed method. Finally, Sect. 6.6 concludes the chapter.

6.2 Related Work

6.2.1 Notations

In this section, we establish the key notations employed throughout this chapter. Matrices are denoted in bold uppercase, vectors in bold lowercase, and scalars in regular lowercase. For a matrix $\mathbf{A}$, the element at the ijth position in the ith row and jth column is represented by A_{ij}. The total number of views is denoted as V and the number of samples as n. The data matrix for view v is expressed as $\mathbf{X}^v = (\mathbf{x}_1^v, \mathbf{x}_2^v, \ldots, \mathbf{x}_n^v) \in \mathbb{R}^{n \times d^v}$, where d^v is the dimension of the feature vector in view v ($v = 1, \ldots, V$). The total number of clusters is denoted as C. The ℓ_2 norm of a matrix $\mathbf{A} \in \mathbb{R}^{n \times m}$ is symbolized by $\|\mathbf{A}\|_2 = \sqrt{\sum_{i=1}^{n} \sum_{j=1}^{m} |A_{ij}|^2}$. The trace and transpose of matrix $\mathbf{A}$ are represented by $Tr(\mathbf{A})$ and $\mathbf{A}^T$, respectively. The main notations employed throughout this chapter are summarized in Table 6.1.

Table 6.1 Some notations used throughout the chapter

Notation	Description
n	Number of samples
C	Number of clusters
V	Number of views
d^v	Dimension of the feature vector of the vth view
$\mathbf{X}^v = (\mathbf{x}_1^v, \mathbf{x}_2^v, \ldots, \mathbf{x}_n^v)$	Data matrix of the vth view $\in \mathbb{R}^{d^v \times n}$
$\mathbf{x}_i^v$	The ith column of $\mathbf{X}^v$
$\mathbf{K}^v$	Kernel matrix of the vth view $\in \mathbb{R}^{n \times n}$
$\|\mathbf{A}\|_2$	ℓ_2 norm of the matrix $\mathbf{A}$
$\mathbf{A}_{i*}$	The ith row of the matrix $\mathbf{A}$
$\mathbf{S}^v$	Similarity matrix of the vth view $\in \mathbb{R}^{n \times n}$
$\mathbf{S}^*$	Consensus similarity matrix $\in \mathbb{R}^{n \times n}$
$\mathbf{L}^v$	Laplacian matrix of the vth view $\in \mathbb{R}^{n \times n}$
$\mathbf{L}^*$	Consensus Laplacian matrix $\in \mathbb{R}^{n \times n}$
$\mathbf{P}^v$	Spectral projection matrix of the vth view $\in \mathbb{R}^{n \times C}$
$\mathbf{P}^*$	Consensus spectral projection matrix of the vth view $\in \mathbb{R}^{n \times C}$
$\mathbf{H}$	Nonnegative embedding matrix $\in \mathbb{R}^{n \times C}$
$\mathbf{I}$	Identity matrix
λ	Balance parameter

6.2.2 Related Work

Due to its straightforward implementation, the spectral clustering method [31] has garnered significant attention in the realm of multi-view clustering. The underlying principle of this method revolves around computing the most pertinent (smallest) eigenvectors of the normalized Laplacian matrix, derived from the similarity matrix between data samples. Subsequently, each row of the matrix formed from the eigenvectors of the Laplacian matrix serves as the representation of a sample. These representations are then clustered into distinct clusters using an independent algorithm, such as k-means.

The co-training approach, as outlined in [17], ensures that samples from different views are placed in the same cluster by adjusting the similarity values of a particular view based on the clustering assignments of the other view. Additionally, the co-regulated algorithm [18], which amalgamates different similarity matrices from various views using an adaptive approach to derive the clustering assignments, is another noteworthy method for spectral clustering. However, in these methods, all views are assigned equal weight. To address this limitation, several weighted multi-view clustering methods, such as those proposed by [30, 35], have been developed. These approaches enhance the algorithm's complexity by introducing hyperparameters to assign weights to each view.

To overcome this challenge, various automatically weighted multi-view clustering techniques [14, 15, 41] have been introduced. These techniques allow for adjusting the weighting of each view without the need for additional parameters. Multi-view subspace-based clustering (MVSC) approaches transform the entire dataset into subspaces to derive the most coherent data representation, which can then be clustered accordingly [1, 33, 34]. Kernel techniques [30] are employed to transform input data into linearly separable data, addressing the nonlinearity of data with diverse shapes.

Matrix factorization techniques [7, 23, 32] represent powerful tools for clustering across multiple views. These methods articulate matrices in a low-dimensional latent space, leading to expedited computation and improved results compared to alternative techniques. In particular, the nonnegative matrix factorization (NMF) approach, as employed in [36], incorporates two constraints: one on the labels of specific samples and the other exploiting the sparsity of representations.

In the realm of nonnegative rank-reduced regression (NRRR), discussed in [8], the authors leverage metric distance learning and clustering, assuming a coherent framework for rank-reduced regression. Further contributions in [12] introduce ensemble clustering by propagating cluster-wise similarities with hierarchical consensus technique (ECPCS-HC) and ensemble clustering by propagating cluster-wise similarities with meta-cluster-based consensus method (ECPCS-MC). These methods employ random walks based on effective cluster-wise similarity propagation, along with additional consensus functions, to optimize clustering results.

The consistency-aware graph-based multi-view clustering (CI-GMVC) approach, proposed in [9], incorporates orthogonality constraints to analyze both consistent and inconsistent components across multiple views. This method partitions the similarity matrix into two distinct graphs, emphasizing the importance of consistency.

A novel method, multi-view clustering in latent embedding space (MCLES), is suggested in [2]. This method jointly generates the latent embedding space and cluster index matrix, uncovering the general structure of the data and exploiting complementary information between multiple views.

The nonnegative embedding and spectral embedding (NESE) method, detailed in [11], employs a nonnegative embedding matrix as a form of convolution between the similarity matrix and spectral representation matrix, determining cluster labels based on this matrix. Another method, multi-view spectral clustering via sparse graph learning (S-MVSC) [10], computes a sparse and joint similarity matrix across all views, demonstrating comparable complexity to single-view spectral clustering and faster performance than some alternative multiview methods.

In [3], a modified version of NESE, called constrained nonnegative embedding and spectral embedding (CNESE), is introduced. This approach enhances the nonnegative embedding matrix by adding two types of constraints while preserving the major benefits of the prior NESE approach.

A distinctive approach, multi-view spectral clustering with self-taught robust graph learning (MCSRGL), is presented in [4]. This method introduces improvements by creating a cluster-label correlation graph and subjecting the cluster membership matrix to a smoothing condition to enhance consistency with both the cluster-label correlation graph and the original datasets' graphs.

Moreover, [5] proposes a method named multi-view clustering via consensus graph learning and nonnegative embedding (MVCGE). This technique simultaneously generates the consensus spectral representation matrix, unified similarity matrix, and nonnegative embedding matrix using kernel matrices from different views as input, eliminating the need for additional steps like k-means clustering or spectral rotation algorithms.

6.3 Proposed Approach

Motivated by various consensus multi-view methods, such as the approach proposed in [38], which simultaneously computes individual similarity matrices for each view, the unified similarity matrix, and final clustering assignments, we introduce a novel method: single-phase multi-view clustering using unified graph learning and spectral representation (MCGLSR). This method primarily involves computing all view-based similarity matrices and the unified similarity matrix to minimize the impact of noisy views. Additionally, MCGLSR calculates the nonnegative embedding matrix, serving as a soft clustering assignment and eliminating the need for post-processing steps. The weight of each view is also computed automatically, requiring no additional parameters, and all these components are calculated simultaneously for an optimal solution.

MCGLSR concurrently determines several elements: (1) individual similarity matrices, (2) consensus similarity matrix, (3) unified spectral representation matrix, and (4) consensus cluster index matrix, all without additional parameters. The method begins with an explanation of its essential components, followed by the global optimization problem. Given $\mathbf{X}^v$, the raw data matrix for each view, expressed as $\mathbf{X}^v = (\mathbf{x}_1^v, \mathbf{x}_2^v, \ldots, \mathbf{x}_n^v) \in \mathbb{R}^{d^v \times n}$, where n is the number of samples and d^v is the dimension of the feature vector in view v ($v = 1, \ldots, V$), MCGLSR aims to group data points into C disjoint clusters. Gaussian kernel matrices are denoted by $\mathbf{K}^v$.

The method simultaneously determines the following matrices: $\mathbf{S}^* \in \mathbb{R}^{n \times n}$, $\mathbf{P}^* \in \mathbb{R}^{n \times C}$, $\mathbf{S}^v \in \mathbb{R}^{n \times n}$, and $\mathbf{H} \in \mathbb{R}^{n \times C}$. Six elements constitute our proposed criterion.

The first and second terms of our method represent the expressive property and the kernel matrix of each view, describing individual similarity matrices. These terms are expressed by the following equation:

$$\min_{\mathbf{S}^v, v=1,\ldots,V} \sum_{v=1}^{V} Tr\,(\mathbf{K}^v - 2\,\mathbf{K}^v\,\mathbf{S}^v + \mathbf{S}^{v\,T}\,\mathbf{K}^v\,\mathbf{S}^v) + ||\mathbf{S}^v||_2^2 \;\; s.t. \;\; 0 \leq \mathbf{S}^v \leq 1, \;\; \mathbf{S}^v\mathbf{1} = \mathbf{1}, \;\; diag(\mathbf{S}^v) = \mathbf{0}. \tag{6.1}$$

The inclusion of the second term aims to prevent trivial solutions.

In the context of multi-view clustering, the objective is to assign each instance to the same cluster across multiple views. As a result, the similarity matrix $\mathbf{S}^v$ for each view is ideally closest to a unified similarity matrix $\mathbf{S}^*$. This approach facilitates the effective learning of distinct and complementary information offered by multiple views while mitigating the impact of noise or outliers within a specific view. Rather than explicitly fusing similarities, we assume that the individual similarity matrices $\mathbf{S}^v$ are proximate to an unknown common similarity matrix $\mathbf{S}^*$. Consequently, our third term is defined as:

$$\min_{\mathbf{S}^*} \sum_{v=1}^{V} \|\mathbf{S}^* - \mathbf{S}^v\|_2^2. \tag{6.2}$$

As previously mentioned, our objective is to directly obtain cluster assignments from a soft cluster assignment matrix, represented by the nonnegative embedding matrix $\mathbf{H}$. This matrix is constructed through the convolution of the unified spectral projection matrix $\mathbf{P}^*$ over its associated unified graph $\mathbf{S}^*$. The cluster affiliation of a data point is indicated by the index of the largest value in the row vector $\mathbf{H}_{i*}$. The fourth term in our optimization problem is expressed as:

$$\min_{\mathbf{H},\mathbf{P}^*} \|\mathbf{S}^* - \mathbf{H}\,\mathbf{P}^{*T}\|_2 \;\; s.t.\,, \;\; \mathbf{H}^T\,\mathbf{H} = \mathbf{I}\,, \;\; \mathbf{P}^{*T}\,\mathbf{P}^* = \mathbf{I}. \tag{6.3}$$

To enhance the clustering results, we introduce a constraint on the smoothness of cluster assignments across the matrix $\mathbf{H}$. This term is expressed as:

$$\min_{\mathbf{H}} \left(\frac{1}{2} \sum_i \sum_j \|\mathbf{H}_{i*} - \mathbf{H}_{j*}\|_2^2\, S_{ij}^* \right) = \min_{\mathbf{H}} Tr\,\left(\mathbf{H}^T\,\mathbf{L}^*\,\mathbf{H}\right), \tag{6.4}$$

where $\mathbf{H}_{i*}$ refers to the ith row of the matrix $\mathbf{H}$, and $\mathbf{H}_{j*}$ refers to the jth row of the same matrix, $\mathbf{L}^* = \mathbf{D}^* - \mathbf{S}^* \in \mathbb{R}^{n\times n}$ represents the consensus Laplacian matrix corresponding to the consensus similarity matrix. The related diagonal matrix $\mathbf{D}$ is defined by $D_{ii} = \sum_{j=1}^{n} \frac{S_{ij}+S_{ji}}{2}$. This term signifies that if we aim to ensure that two samples $\mathbf{x}_i$ and $\mathbf{x}j$ belong to the same cluster, their similarity value should be substantial, and the values of $\mathbf{H}i*$ and $\mathbf{H}_{j*}$ should be similar.

The final term in our objective function asserts that the graph should closely resemble a graph with exactly C connected components or clusters. Once the graph is defined, this term is minimized to produce a spectral embedding. The corresponding optimization problem is formulated as follows:

$$\min_{\mathbf{P}^*} Tr\,(\mathbf{P}^{*T}\,\mathbf{L}^*\,\mathbf{P}^*) \;\; s.t. \;\; \mathbf{P}^T\,\mathbf{P} = \mathbf{I}. \tag{6.5}$$

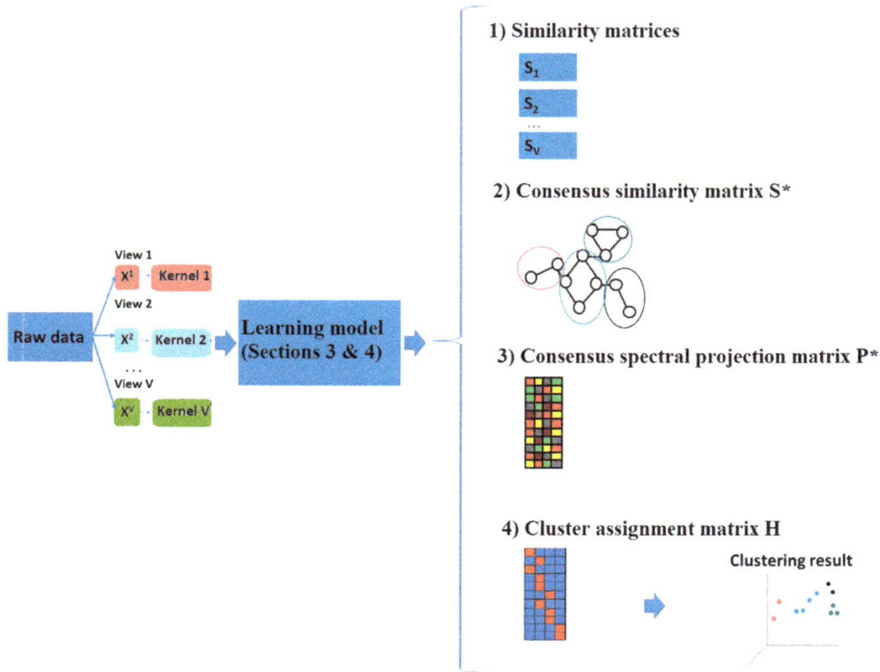

Fig. 6.1 Illustration of the MCGLSR method

Therefore, our comprehensive optimization problem arises from the amalgamation of all the aforementioned terms. The final objective function is given by Eq. (6.6):

$$\min_{\mathbf{S}^v, \mathbf{S}^*, \mathbf{P}^*, \mathbf{H}} \sum_{v=1}^{V} \left\{ Tr\,(\mathbf{K}^v - 2\,\mathbf{K}^v\,\mathbf{S}^v + \mathbf{S}^{vT}\,\mathbf{K}^v\,\mathbf{S}^v) + ||\mathbf{S}^v||_2^2 + \lambda_1\,||\mathbf{S}^* - \mathbf{S}^v||_2^2 \right\}$$
$$+ \lambda_2\,||\mathbf{S}^* - \mathbf{H}\,\mathbf{P}^{*T}||_2^2 + \lambda_3\,Tr\,(\mathbf{H}^T\,\mathbf{L}^*\,\mathbf{H}), + \lambda_4\,Tr\,(\mathbf{P}^{*T}\,\mathbf{L}^*\,\mathbf{P}^*), \tag{6.6}$$

The optimization problem is governed by regularization parameters λ_1, λ_2, λ_3, and λ_4.

Figure 6.1 provides a visual representation of the proposed MCGLSR approach.

6.4 Optimization of the Proposed MCGLSR (Eq. (6.6))

This section outlines the optimization scheme for the objective function described in (6.6). To address our optimization problem, we propose an effective iterative algorithm based on an alternating minimization scheme. This scheme involves updating a designated matrix while keeping the other unknown matrices fixed. The optimization procedure for MCGLSR is introduced to update the matrices $\mathbf{S}^v$, $\mathbf{S}^*$, $\mathbf{P}^*$, and $\mathbf{H}$. The matrices $\mathbf{S}^v$ and $\mathbf{P}^v$ are initialized using the procedure outlined in [24]. As for $\mathbf{S}^*$ and $\mathbf{P}^*$, they are initialized by taking the average of the obtained matrices $\mathbf{S}^v$ and $\mathbf{P}^v$, respectively.

Therefore, the algorithm executes the following update steps iteratively.

Update H Given $\mathbf{S}^v$, $\mathbf{S}^*$, and $\mathbf{P}^*$, the nonnegative embedding matrix $\mathbf{H}$ is adjusted by computing the derivative of the function in (6.6) with respect to $\mathbf{H}$:

$$\frac{\partial f}{\partial \mathbf{H}} = 2\,\lambda_2\,(\mathbf{H} - \mathbf{S}^*\,\mathbf{P}^*\,) + 2\,\lambda_3\,\mathbf{L}^*\,\mathbf{H}\,. \tag{6.7}$$

Setting this derivative to zero yields the value of $\mathbf{H}$, as given by Eq. (6.8).

$$\mathbf{H} = \left(\lambda_2\,\mathbf{I} + \lambda_3\,\mathbf{L}^*\right)^{-1}\,\left(\lambda_2\,\mathbf{S}^*\,\mathbf{P}^*\right). \tag{6.8}$$

After calculating $\mathbf{H}$ using Eq. (6.8), an orthogonalization step is applied to the obtained matrix. Additionally, the positivity condition is ensured by applying the widely used element-wise rectified linear unit (ReLU) operator to the components of the resulting matrix $\mathbf{H}$.

Update P* Given $\mathbf{S}^v$, $\mathbf{S}^*$, and $\mathbf{H}$, the objective function of our method will be:

$$\min_{\mathbf{P}^*}\; Tr(\mathbf{P}^{*T}\,\mathbf{L}^*\,\mathbf{P}^*) + \frac{\lambda_2}{\lambda_4}\,\left\|\mathbf{S}^* \;- \mathbf{H}\,\mathbf{P}^{*T}\right\|_2^2 \;\; s.t.\,\mathbf{P}^{*T}\,\mathbf{P}^* = \mathbf{I}. \tag{6.9}$$

To solve Eq. (6.9), the following equation is used:

$$\left\|\mathbf{S}^*\,\mathbf{P}^* \;- \mathbf{H}\,\right\|_2^2 = Tr[(\mathbf{S}^*\,\mathbf{P}^* \;- \mathbf{H}\,)^T\,(\mathbf{S}^*\,\mathbf{P}^* \;- \mathbf{H}\,)]. \tag{6.10}$$

The derivative of the functional in (6.9) with respect to $\mathbf{P}^*$ is given by:

$$\frac{\partial f}{\partial \mathbf{P}^*} = 2\,\mathbf{L}^*\,\mathbf{P}^* \;+ 2\,\frac{\lambda_2}{\lambda_4}\,\mathbf{S}^{*T}\,\mathbf{S}^*\,\mathbf{P}^* \;- 2\frac{\lambda_2}{\lambda_4}\,\mathbf{S}^{*T}\,\mathbf{H}. \tag{6.11}$$

To obtain the optimal $\mathbf{P}^*$, setting this derivative to zero yields:

$$\mathbf{P}^* = \left(\mathbf{L}^* + \frac{\lambda_2}{\lambda_4}\,\mathbf{S}^{*T}\,\mathbf{S}^*\right)^{-1}\,\frac{\lambda_2}{\lambda_4}\,\mathbf{S}^{*T}\,\mathbf{H}. \tag{6.12}$$

To fulfill the orthogonality condition, an orthogonalization step is applied to the obtained $\mathbf{P}^*$.

Update $\mathbf{S}^v$ When $\mathbf{S}^*$, $\mathbf{P}^*$, and $\mathbf{H}$ are fixed, the problem to be solved is described below:

$$\min_{\mathbf{S}^v} \sum_{v=1}^{V} \left(Tr\,(\mathbf{K}^v - 2\,\mathbf{K}^v\,\mathbf{S}^v + \mathbf{S}^{vT}\,\mathbf{K}^v\,\mathbf{S}^v) + ||\mathbf{S}^v||_2^2 + \lambda_1\,||\mathbf{S}^* - \mathbf{S}^v||_2^2 \right) s.t.\ 0$$
$$\leq \mathbf{S}^v \leq 1,\ (\mathbf{S}^v)^T\mathbf{1} = \mathbf{1},\ diag(\mathbf{S}^v) = \mathbf{0}. \tag{6.13}$$

Setting the derivative of this equation to zero, the expression $\hat{\mathbf{S}^v}$, which is the similarity matrix for each view without considering the constraints, is obtained by the following equation:

$$\hat{\mathbf{S}^v} = \left(\mathbf{K}^v + (1+\lambda_1)\,\mathbf{I}\right)^{-1} \left(\mathbf{K}^v + \lambda_1\,\mathbf{S}^*\right). \tag{6.14}$$

To adhere to the conditions for each similarity matrix, it is necessary to project it into a constrained space. Furthermore, for a fixed v, the optimization of each row $\mathbf{S}^v_{i,:}$ is independent of the optimization of the other rows.

Therefore, the solution for $\mathbf{S}^v$ involves addressing the following problem: for each view v and for each row of the similarity matrix, the following minimization problem is obtained:

$$\min_{0 \leq \mathbf{S}^v_{i,:} \leq 1,\ \mathbf{S}^{vT}_{i,:}\mathbf{1}=1,\ S^v_{i,i}=0} ||\mathbf{S}^v_{i,:} - \hat{\mathbf{S}^v_{i,:}}||_2^2. \tag{6.15}$$

The Lagrangian function of the above minimization problem can be written as follows:

$$\mathcal{L}(\mathbf{S}^v_{i,:}, \alpha^v_i, \beta_i) = ||\mathbf{S}^v_{i,:} - \hat{\mathbf{S}^v_{i,:}}||_2^2 - \alpha^v_i(\mathbf{S}^{vT}_{i,:}\mathbf{1} - 1) - \beta_i^T\,\mathbf{S}^v_{i,:}, \tag{6.16}$$

Here, α^v_i and $\beta^v_i \geq \mathbf{0}$ are Lagrange multipliers. Consequently, following the KKT conditions, the ith row of the similarity matrix for each view $\mathbf{S}^v$ is determined as [25, 40]:

$$\mathbf{S}^v_{i,:} = \max(\hat{\mathbf{S}^v_{i,:}} + \alpha^v_i\mathbf{1}^T, 0),\ S^v_{i,i} = 0. \tag{6.17}$$

The elements of $\hat{\mathbf{S}^v_{i,:}}$ are rearranged in descending order, yielding $\mathbf{S}'^v_{i,:} = [S'^v_{i,1}, \ldots, S'^v_{i,n}]^T$. To satisfy the constraint $\mathbf{S}^{vT}_{i,:}\mathbf{1} = 1$ and assume the use of K-nearest neighbors to represent each data point (i.e., the vector $\mathbf{S}^v_{i,:}$ contains only K non-zero elements), we obtain $\alpha_i = \frac{1}{K} - \frac{1}{K}\sum_{l=1}^{K} S'^v_{i,l}$. The closed-form solution for $\mathbf{S}^v_{i,:}$ is given by:

$$S^v_{i,j} = S'^v_{i,m} + \frac{1}{K} - \frac{1}{K}\sum_{l=1}^{K} S'^v_{i,l} \quad \text{if} j \in \mathcal{N}_k(i);\quad \text{otherwise } S^v_{i,j} = 0. \tag{6.18}$$

where $m \in \{1, 2, \ldots, K\}$ is the corresponding index of the jth element in the reordered vector $\mathbf{S}'^{v}_{i,:}$.

Update $\mathbf{S}^*$ With $\mathbf{S}^v$, $\mathbf{P}^*$, and $\mathbf{H}$ fixed, the problem to be addressed is described below:

$$\min_{\mathbf{S}^*} \sum_{v=1}^{V} \lambda_1 ||\mathbf{S}^* - \mathbf{S}^v||_2^2 + \lambda_2 ||\mathbf{S}^* - \mathbf{H}\,\mathbf{P}^{*T}||_2^2 + \lambda_3\, Tr\,(\mathbf{H}^T\,\mathbf{L}^*\,\mathbf{H})$$
$$+ \lambda_4\, Tr\,(\mathbf{P}^{*T}\,\mathbf{L}^*\,\mathbf{P}^*). \tag{6.19}$$

The following identity is a known result derived from spectral clustering analysis:

$$Tr\,(\mathbf{U}^T\,\mathbf{L}^v\,\mathbf{U}) = \frac{1}{2}\sum_i\sum_j \left\|\mathbf{U}_{i*} - \mathbf{U}_{j*}\right\|^2 S_{ij}^v = \frac{1}{2}\,Tr\,(\mathbf{Q}\,\mathbf{S}^v), \tag{6.20}$$

Here, $\mathbf{U}_{i*}$ and $\mathbf{U}_{j*}$ are the ith and jth rows of the matrix $\mathbf{U}$, and $Q_{ij} = Q_{ji} = \left\|\mathbf{U}_{i*} - \mathbf{U}_{j*}\right\|^2$.

Leveraging this identity, Eq. (6.19) can be expressed as follows:

$$\min_{\mathbf{S}^*} \sum_{v=1}^{V} \lambda_1 ||\mathbf{S}^* - \mathbf{S}^v||_2^2 + \lambda_2 ||\mathbf{S}^* - \mathbf{H}\,\mathbf{P}^{*T}||_2^2 + \frac{\lambda_3}{2}\, Tr\,(\mathbf{Q}_H\,\mathbf{S}^*)$$
$$+ \frac{\lambda_4}{2}\, Tr\,(\mathbf{Q}_P\,\mathbf{S}^*), \tag{6.21}$$

Here, the matrices $\mathbf{Q}_H$ and $\mathbf{Q}_P$ are the pairwise distance matrices associated with the rows of matrices $\mathbf{H}$ and $\mathbf{P}^*$, respectively.

Setting the derivative of Eq. (6.21) with respect to $\mathbf{S}^*$ to zero yields the following solution:

$$\mathbf{S}^* = ReLU\left\{\left(\left(\sum_{v=1}^{V} \lambda_1 + \lambda_2\right)\mathbf{I}\right)^{-1}\left(\sum_{v=1}^{V} \lambda_1\,\mathbf{S}^v + \lambda_2\,\mathbf{H}\,\mathbf{P}^{*T}\right.\right.$$
$$\left.\left.-\frac{1}{2}\lambda_3\,\mathbf{Q}_H - \frac{1}{2}\lambda_4\,\mathbf{Q}_P\right)\right\}. \tag{6.22}$$

Hence, the KNN algorithm is employed for each row of the matrix $\mathbf{S}^*$ to identify the K most similar samples for each given data point.

Algorithm 1: Steps of the proposed MCGLSR technique.

Algorithm 1 MCGLSR

Input:	Data matrices $\mathbf{X}^v \in \mathbb{R}^{n \times d^v}$, $v = 1, \ldots, V$, and their related kernel matrices $\mathbf{K}^v$ Parameters λ_1, λ_2, λ_3, and λ_4
Output:	The consensus nonnegative embedding matrix $\mathbf{H}$ The individual similarity graphs $\mathbf{S}^v$ The unified similarity matrix $\mathbf{S}^*$ The unified spectral projection matrix $\mathbf{P}^*$
Initialization:	Initialize the similarity matrices of all views $\mathbf{S}^v$ and their related spectral projection matrices $\mathbf{P}^v$. Initialize $\mathbf{S}^*$ and its corresponding $\mathbf{P}^*$ as mentioned before. **Repeat** Update $\mathbf{H}$ using Eq. (6.8). Update $\mathbf{P}^*$ using Eq. (6.12). Update $\mathbf{S}^v$, $v = 1, \ldots, V$ using Eq. (6.14). Update $\mathbf{S}^*$ using Eq. (6.22). **Until H does not change or the maximum number of iterations is reached**

6.4.1 Computational Complexity

In this section, we analyze the computational complexity of the proposed MCGLSR method, which involves four main steps: updating $\mathbf{H}$, $\mathbf{P}^*$, $\mathbf{S}^v$, and $\mathbf{S}^*$ (see Algorithm 1).

The computation of the V kernel matrices has a computational cost of $O(2n^2k)$, where k is the sum of the dimensions of the instances in the V views ($k = d^1 + d^2 + \ldots + d^V$).

To obtain the matrices $\mathbf{H}$ and $\mathbf{P}^*$ (steps 1 and 2), a matrix inversion of an $n \times n$ matrix is required (or solving a linear system whose square matrix size is $n \times n$). Thus, the computational cost for the first and second steps is $O(n^3)$. If the orthogonalization of the matrices $\mathbf{H}$ and $\mathbf{P}^*$ is invoked, we should take into account the associated cost, which is $O(nC^2)$, where C is the number of clusters.

To estimate the graph matrix of each view $\mathbf{S}^v$ (step 3) and the unified similarity matrix $\mathbf{S}^*$ (step 4), a matrix inversion of size $n \times n$ is required. Thus, the computational cost for the third and fourth steps is $O((V+1)n^3)$.

Let τ be the number of iterations of the proposed iterative algorithm. The total computational complexity of the proposed method is then $O\left(n^2k + \tau(2nC^2 + (V+3)n^3)\right) \approx O\left(n^2k + \tau(2nC^2 + n^3)\right)$.

Although the proposed multi-view clustering method may not be the fastest in terms of computation time, its computational complexity is comparable to many other existing graph-based multi-view clustering methods.

Table 6.2 Description of the datasets used in the chapter

View	ORL	COIL20	BBCSport	MSRCv1	MNIST-25000
1	(512) GIST	(1024) Intensity	(3183) View1	(512) GIST	(4096) VGG16 FC1
2	(59) LBP	(3304) LBP	(3203) View2	(256) LBP	(2048) Resnet50
3	(864) HOG	(6750) Gabor	–	(24) Color moment	–
4	(254) Centrist	–	–	(254) Centrist	–
5	–	–	–	(512) SIFT	–
# of samples	1440	400	544	210	25,000
# of classes	20	40	5	7	10

6.5 Performance Evaluation

6.5.1 Experimental Setup

In this section, we evaluate the effectiveness of our proposed approach using four small datasets: three image datasets (ORL,[1] COIL20,[2] and MSRCv1[3]), and one text dataset "BBCSport.[4]" Additionally, we utilize the MNIST-25000[5] dataset, which can be considered a large dataset in the field of graph-based multi-view clustering.

The description of the five datasets used in this study can be found in Table 6.2. In this table, the number between the parentheses indicates the dimension of the feature vector of each view. To ensure a fair comparison, experimental results were obtained using codes requested from the corresponding authors and implemented with wide parameter ranges. The code for our proposed method will be available at the following link: https://github.com/SallyHajjar/MCGLSR.git.

Compared Methods

Various relevant state-of-the-art approaches are compared with our proposed approach, including the auto-weighted multi-view clustering with single kernel (MVCSK) [13], the nonnegative embedding and spectral embedding method (NESE) [11], the sparse multi-view spectral clustering via graph learning method (S-MVSC) [10], the consistency-aware and Inconsistency-aware graph-based multi-view clustering (CI-GMVC) presented in [9], the approach called multi-view clustering in latent embedding space (MCLES) in [2], the multi-view spectral clustering via constrained nonnegative embedding (CNESE) presented in [3],

[1] https://cam-orl.co.uk/facedatabase.html

[2] http://www.cs.columbia.edu/CAVE/software/softlib/coil-20.php

[3] https://www.researchgate.net/publication/335857675

[4] http://mlg.ucd.ie/datasets/segment.html

[5] https://www.kaggle.com/datasets/hojjatk/mnist-dataset

the multi-view spectral clustering via integrating label and data graph learning (MSLDGL) [6], and multi-view spectral clustering with a self-taught robust graph learning (MCSRGL) [4]. Additionally, spectral clustering best (SC-Best) [31], which implements the spectral clustering algorithm for each view separately and then reports the best result for the best view, is also used as a competing method.

In our approach, the initialization of the matrices $\mathbf{S}^v$ and $\mathbf{P}^v$ follows the same procedure as in [11], and the matrices $\mathbf{S}^*$ and $\mathbf{P}^*$ are initialized as mentioned before.

Parameter Setting

According to our algorithm, there are four main parameters to be determined: λ_1, λ_2, λ_3, and λ_4. The value of λ_1 is chosen from the set $\{0.0005, 0.05, 0.5, 1, 10, 10^{+2}\}$, the value of λ_2 is chosen from the set $\{10^{-7}, 10^{-6}, 10^{-5}, 10^{-4}, 10^{-3}, 10^{-2}, 0.1, 0.2, 0.3, 0.4, 0.5, 0.6, 0.7\}$, the value of λ_3 is chosen from the set $\{0.005, 0.05, 0.1, 0.2, 0.3, 0.4, 0.5, 1, 10, 10^{+2}\}$, and the value of λ_4 is chosen from the set $\{10^{-7}, 10^{-6}, 10^{-5}, 10^{-4}, 10^{-3}, 10^{-2}, 0.1, 1, 10, 10^{+2}\}$.

To determine the best values for the parameters λ_1, λ_2, λ_3, and λ_4, we used a grid search procedure. This strategy is widely used in the literature to select the best values for regularization parameters in conjunction with unsupervised clustering methods. In our experiments, for all datasets used, we defined a set of values for each parameter (see the sets above). This set of values was selected based on the relevant literature and our preliminary experiments. Once each parameter has a set of values, we have a total set of parameter combinations, each of which is used by the optimization scheme. The total number of these combinations is equal to the product of the numbers of values of each balance parameter. The best combination is then the one that gives the best evaluation metric. In brief, in our work, the search grid was the same for all datasets used. In other words, the tested values were the same for all datasets. However, we emphasize that the optimal values of the parameters may depend on the dataset used.

We incorporate a Gaussian kernel in our approach, defined by the entry $K_{ij} = \exp\left(-\frac{\|\mathbf{x}_i - \mathbf{x}_j\|^2}{2^{T_0}\,\sigma_0^2}\right)$ for a pair of data samples $\mathbf{x}_i$ and $\mathbf{x}_j$. In typical learning algorithms, σ_0^2 is set to the average squared distance between sample pairs within the considered view. The parameter T_0 is a user-defined integer regulating the kernel's scale, and our experiments demonstrate that setting T_0 to 2 yields optimal results across all datasets.

Additionally, our method involves determining the K closest samples for a given data point. The value of K signifies the number of nonzero elements in each row of the similarity matrices for all views and the consensus similarity matrix. We choose this value in the range of 5–20. A thorough exploration of the sensitivity of this parameter is presented in Sect. 6.5.3 of our study. In our study, we employ four widely recognized evaluation measures for clustering: clustering accuracy (ACC), normalized mutual information (NMI), purity, and adjusted Rand index (ARI) [37]. Higher values for these parameters indicate better results.

6.5.2 Experimental Results

The experimental outcomes are detailed in Tables 6.3 and 6.4. Our MCGLSR method is compared with other competing methods in these tables. The superior results are highlighted in bold, and standard deviations, denoted in parentheses, depict the variation across multiple experiments.

For two-stage clustering approaches, involving data representation followed by clustering, performance is assessed by repeating clustering multiple times and reporting average and standard deviation for ACC, NMI, purity, and ARI. These results are derived from a multi-view clustering paper [11], where different methods were executed 30 times for each two-stage clustering method. For direct clustering approaches, the standard deviation remains zero.

Table 6.3 showcases the outcomes for the ORL, COIL20, BBCSport, and MSRCv1 datasets. It is evident from this table that our proposed approach consistently outperforms other methods across all datasets. Table 6.4 provides a comparison between our method and state-of-the-art methods, namely MVCSK, NESE, MCLES, and MCSRGL, applied to the MNIST-25000 dataset.

6.5.3 Parameter Sensitivity

As outlined in Algorithm 1, our objective function involves four explicit parameters: λ_1, λ_2, λ_3, and λ_4. Additionally, the parameter K used during algorithm iterations requires fine-tuning. Extensive studies reveal that the optimal value for the parameter T_0, determining the Gaussian kernel's scale, is set to 2. Accordingly, we fix T_0 at 2. The sensitivity of these parameters is thoroughly examined using the MSRCv1 dataset.

In Fig. 6.2, the impact of λ_2 and λ_3 on clustering indicators ACC and NMI is depicted for the MSRCv1 dataset, with λ_1, λ_4, and K held constant. The optimal values for ACC and NMI are observed when λ_2 is 10^{-5} and λ_3 is 0.5.

Furthermore, Fig. 6.3 illustrates the influence of λ_1 (left) and λ_4 (right) on clustering results, with the other three parameters set to their optimal values for the MSRCv1 dataset. Optimal cluster indicator performance (ACC and NMI) is achieved with $\lambda_1 = 1$ and $\lambda_4 = 10$.

The effect of parameter K on clustering results is analyzed in Fig. 6.4, keeping the other four parameters fixed at their optimal values. The highest values for cluster indicators ACC and NMI using the MCGLSR method for the MSRCv1 dataset are observed when K is set to 17.

Table 6.3 Clustering results on the ORL, COIL20, BBCSport, and MSRCv1 datasets

Dataset	Method	ACC	NMI	Purity	ARI
ORL	SC-Best [39]	0.66 (± 0.02)	0.76 (± 0.02)	0.71 (± 0.02)	0.67 (± 0.01)
	MVCSK [13]	0.85 (± 0.02)	0.94 (± 0.01)	0.88 (± 0.02)	0.81 (± 0.02)
	NESE [11]	0.82 (± 0.00)	0.91 (± 0.00)	0.85 (± 0.00)	0.75 (± 0.00)
	S-MVSC [10]	0.80 (± 0.02)	0.93 (± 0.01)	0.82 (± 0.02)	0.89 (± 0.01)
	CI-GMVC [9]	0.81 (± 0.00)	0.92 (± 0.00)	0.85 (± 0.00)	0.74 (± 0.00)
	MCLES [2]	0.84 (± 0.00)	0.94 (± 0.00)	0.88 (± 0.00)	0.79 (± 0.00)
	CNESE [3]	0.87 (± 0.00)	0.95 (± 0.00)	0.89 (± 0.00)	0.84 (± 0.00)
	MSLDGL [6]	0.90 (± 0.00)	0.95 (± 0.00)	0.91 (± 0.00)	0.86 (± 0.00)
	MCSRGL [4]	**0.92 (± 0.00)**	**0.96 (± 0.00)**	**0.93 (± 0.00)**	**0.88 (± 0.00)**
	MCGLSR	0.90 (± 0.00)	**0.96 (± 0.00)**	0.90 (± 0.00)	**0.88 (± 0.00)**
COIL20	SC-Best [39]	0.73 (± 0.01)	0.82 (± 0.01)	0.75 (± 0.01)	0.68 (± 0.02)
	MVCSK [13]	0.65 (± 0.04)	0.80 (± 0.02)	0.70 (± 0.03)	0.61 (± 0.05)
	NESE [11]	0.77 (± 0.00)	0.88 (± 0.00)	0.82 (± 0.00)	0.69 (± 0.00)
	S-MVSC [10]	0.62 (± 0.01)	0.86 (± 0.02)	0.77 (± 0.02)	0.97 (± 0.02)
	CI-GMVC [9]	0.86 (± 0.00)	0.94 (± 0.00)	0.90 (± 0.00)	0.83 (± 0.00)
	MCLES [2]	0.79 (± 0.00)	0.88 (± 0.00)	0.83 (± 0.00)	0.75 (± 0.00)
	CNESE [3]	0.82 (± 0.00)	0.88 (± 0.00)	0.82 (± 0.00)	0.78 (± 0.00)
	MSLDGL [6]	0.76 (± 0.00)	0.85 (± 0.00)	0.77 (± 0.00)	0.70 (± 0.00)
	MCSRGL [4]	0.89 (± 0.00)	0.94 (± 0.00)	0.89 (± 0.00)	0.84 (± 0.00)
	MCGLSR	**0.97 (± 0.00)**	**0.97 (± 0.00)**	**0.97 (± 0.00)**	**0.94 (± 0.00)**
	SC-Best [39]	0.72 (± 0.06)	0.60 (± 0.04)	0.72 (± 0.04)	0.48 (± 0.00)
	MVCSK [13]	0.90 (± 0.07)	0.82 (± 0.02)	0.90 (± 0.02)	0.85 (± 0.07)
BBCSport	NESE [11]	0.72 (± 0.00)	0.69 (± 0.00)	0.75 (± 0.00)	0.60 (± 0.00)
	S-MVSC [10]	0.58 (± 0.07)	0.67 (± 0.01)	0.73 (± 0.02)	0.83 (± 0.04)
	CI-GMVC [9]	0.61 (± 0.00)	0.46 (± 0.00)	0.63 (± 0.00)	0.36 (± 0.00)
	MCLES [2]	0.88 (± 0.00)	0.80 (± 0.00)	0.88 (± 0.00)	0.83 (± 0.00)
	CNESE [3]	0.72 (± 0.00)	0.68 (± 0.00)	0.76 (± 0.00)	0.60 (± 0.00)
	MSLDGL [6]	0.75 (± 0.00)	0.70 (± 0.00)	0.80 (± 0.00)	0.63 (± 0.00)
	MCSRGL [4]	0.77 (± 0.00)	0.73 (± 0.00)	0.82 (± 0.00)	0.67 (± 0.00)
	MCGLSR	**0.99 (± 0.00)**	**0.95 (± 0.00)**	**0.99 (± 0.00)**	**0.97 (± 0.00)**
	SC-Best [39]	0.77 (± 0.00)	0.70 (± 0.00)	0.79 (± 0.00)	0.61 (± 0.00)
	MVCSK [13]	0.70 (± 0.02)	0.59 (± 0.03)	0.70 (± 0.02)	0.50 (± 0.04)
MSRCv1	NESE [11]	0.77 (± 0.00)	0.72 (± 0.00)	0.80 (± 0.00)	0.64 (± 0.00)
	S-MVSC [10]	0.60 (± 0.00)	0.69 (± 0.02)	0.74 (± 0.02)	0.79 (± 0.01)
	CI-GMVC [9]	0.74 (± 0.00)	0.72 (± 0.00)	0.77 (± 0.00)	0.59 (± 0.00)
	MCLES [2]	0.90 (± 0.01)	0.83 (± 0.02)	0.90 (± 0.01)	0.77 (± 0.00)
	CNESE [3]	0.86 (±0.00)	0.76 (±0.00)	0.86 (±0.00)	0.72 (±0.00)
	MSLDGL [6]	0.90 (±0.00)	0.81 (±0.00)	0.90 (±0.00)	0.77 (±0.00)
	MCSRGL [4]	0.91 (±0.00)	0.81 (±0.00)	0.92 (±0.00)	0.79 (±0.00)
	MCGLSR	**0.96 (± 0.00)**	**0.92 (± 0.00)**	**0.96 (± 0.00)**	**0.91 (± 0.00)**

Table 6.4 Clustering performance on the MNIST-25000 dataset

Dataset	Method	ACC	NMI	Purity	ARI
MNIST-25000	MVCSK [13]	0.47 (± 0.00)	0.38 (± 0.00)	0.52 (± 0.00)	0.25 (± 0.00)
	NESE [11]	0.72 (± 0.00)	0.75 (± 0.00)	0.77 (± 0.00)	0.65 (± 0.00)
	MCLES [2]	0.73 (± 0.00)	0.76 (± 0.00)	0.78 (± 0.00)	0.64 (± 0.00)
	MCSRGL [4]	0.77 (± 0.00)	0.80 (± 0.00)	0.82 (± 0.00)	0.69 (± 0.00)
	MCGLSR	**0.79 (± 0.00)**	**0.88 (± 0.00)**	**0.84 (± 0.00)**	**0.75 (± 0.00)**

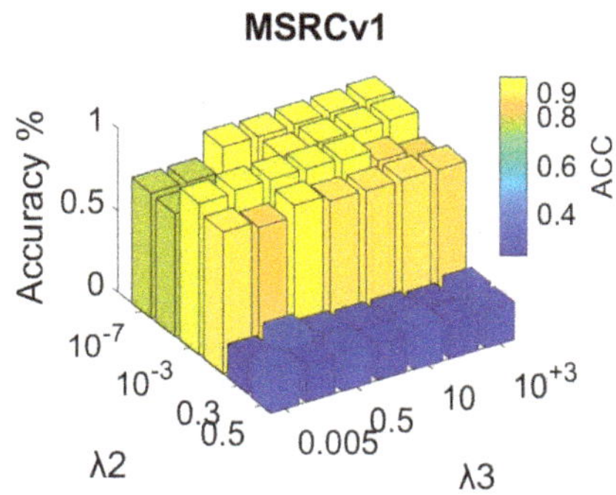

(a)

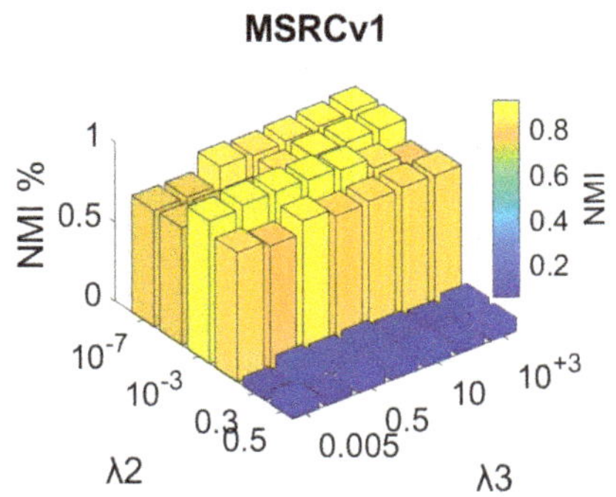

(b)

Fig. 6.2 Clustering results ACC and NMI of the MCGLSR method as a function of λ_2 and λ_3 on the MSRCv1 dataset

6.5.4 *Analysis of Results and Method Comparison*

In our comparative analysis with competing methods, the performance of the well-known spectral clustering algorithm, applied individually to each view and denoted as "SC - Best," typically yields inferior results compared to other multi-view clustering methods, as highlighted in Table 6.3.

The comprehensive results presented in Table 6.3 demonstrate that our proposed method consistently outperforms all competing methods across various datasets, except for the ORL dataset where it falls slightly below the performance of the

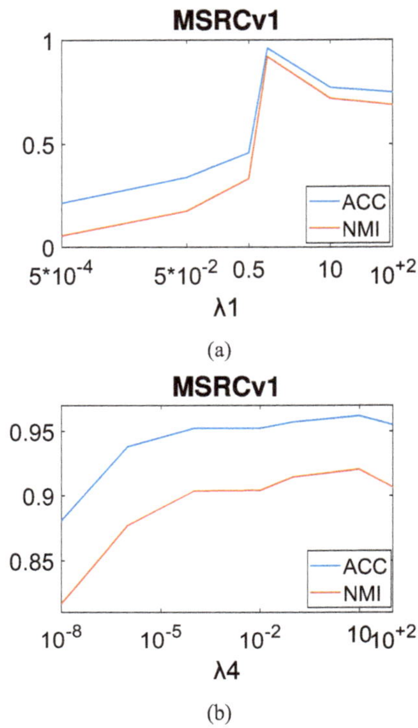

Fig. 6.3 Clustering results ACC and NMI as a function of λ_1 (a) and λ_4 (b) on the MSRCv1 dataset

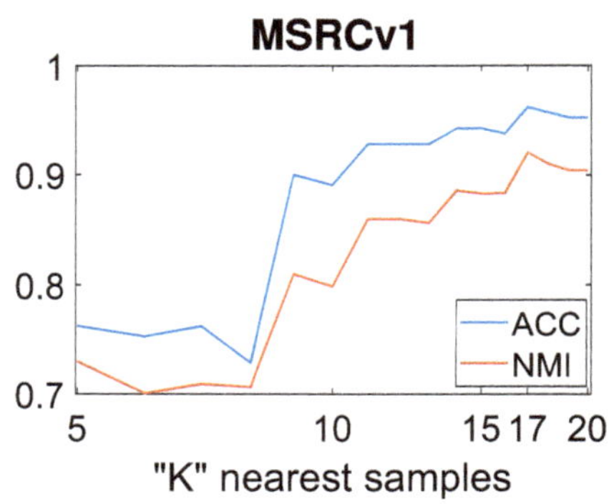

Fig. 6.4 Clustering results ACC and NMI as a function of K on the MSRCv1 dataset

MCSRGL method. Nevertheless, our method's performance remains superior to that of other competing methods.

Additionally, in Table 6.4, the results of applying MCGLSR to the MNIST-25000 dataset are showcased. These results surpass those of the other four methods, indicating the effectiveness of our approach, particularly when applied to large datasets.

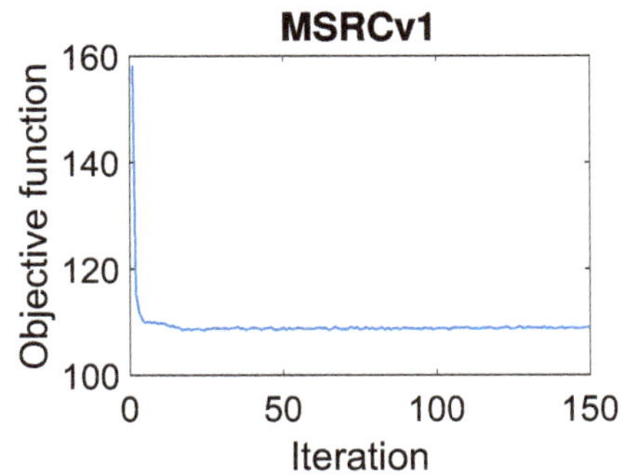

Fig. 6.5 Convergence of our proposed MCGLSR method on the MSRCv1 dataset

6.5.5 Convergence Study

In this section, we delve into the convergence analysis of our proposed method, MCGLSR, focusing on the MSRCv1 dataset. The maximum number of iterations for our method is set at 150. As illustrated in Fig. 6.5, the plot depicts the evolution of the objective function value concerning the number of iterations. Remarkably, our method exhibits a stable and robust convergence property, reaching convergence in less than 20 iterations.

The strong convergence observed in Fig. 6.5 underscores the effectiveness of our approach. Consequently, setting the maximum number of iterations to 150 ensures convergence. Additionally, our algorithm incorporates a convergence criterion based on changes in cluster assignments between iterations. Specifically, the algorithm halts when the change in cluster assignments falls below a predetermined threshold or when the maximum iteration limit is reached. This dual convergence mechanism adds further assurance to the stability and reliability of the MCGLSR approach.

6.6 Conclusion

This study introduces a pioneering approach to tackle the complexities of multi-view clustering, presenting an innovative method that concurrently computes critical components, including individual similarity matrices, a unified similarity matrix, a unified spectral representation matrix, and a nonnegative embedding matrix.

A key feature of our approach lies in its simultaneous computation of these matrices, providing a comprehensive insight into the underlying data structure across multiple views. Additionally, the integration of a unified graph learning process and spectral representation enhances the efficiency and coherence of clustering results.

A notable aspect of this study is the emphasis on refining clustering outcomes through the strategic application of various constraints on the nonnegative embedding matrix. These constraints play a crucial role in optimizing the clustering process, leading to more robust and meaningful results.

To assess the practical effectiveness of our proposed method, we conducted a comprehensive set of experiments using five diverse multi-view real datasets, spanning variations in both size and data types. The obtained results undeniably establish the superior performance of our approach when compared to existing state-of-the-art multi-view clustering methods. Our method not only exhibits its adeptness in effectively handling diverse datasets but also marks a noteworthy advancement in the field by consistently achieving more accurate and reliable clustering outcomes.

In conclusion, the single-phase multi-view clustering using unified graph learning and spectral representation method not only provides a comprehensive solution to the challenges of multi-view clustering but also marks a promising avenue for future research in enhancing the understanding of complex, multifaceted datasets.

References

1. Cao, X., Zhang, C., Fu, H., Liu, S., Zhang, H.: Diversity-induced multi-view subspace clustering. In: Proceedings of the IEEE Conference on Computer Vision and Pattern Recognition, pp. 586–594 (2015)
2. Chen, M.S., Huang, L., Wang, C.D., Huang, D.: Multi-view clustering in latent embedding space. In: Proceedings of the AAAI Conference on Artificial Intelligence, vol. 34, pp. 3513–3520 (2020)
3. El Hajjar, S., Dornaika, F., Abdallah, F.: Multi-view spectral clustering via constrained nonnegative embedding. Inform. Fusion **78**, 209–217 (2022)
4. El Hajjar, S., Dornaika, F., Abdallah, F.: One-step multi-view spectral clustering with cluster label correlation graph. Inform. Sci. **592**, 97–111 (2022)
5. El Hajjar, S., Dornaika, F., Abdallah, F., Barrena, N.: Consensus graph and spectral representation for one-step multi-view kernel based clustering. Knowl.-Based Syst. **241**, 108250 (2022)
6. El Hajjar, S., Dornaika, F., Abdallah, F., Omrani, H.: Multi-view spectral clustering via integrating label and data graph learning. In: International Conference on Image Analysis and Processing, pp. 109–120. Springer, Berlin (2022)
7. Greene, D., Cunningham, P.: A matrix factorization approach for integrating multiple data views. In: Joint European Conference on Machine Learning and Knowledge Discovery in Databases, pp. 423–438. Springer, Berlin (2009)
8. Guo, W., Shi, Y., Wang, S.: A unified scheme for distance metric learning and clustering via rank-reduced regression. IEEE Trans. Syst. Man Cybern. Syst. **51**(8), 5218–5229 (2019)
9. Horie, M., Kasai, H.: Consistency-aware and inconsistency-aware graph-based multi-view clustering. In: 2020 28th European Signal Processing Conference (EUSIPCO), pp. 1472–1476. IEEE, Piscataway (2021)
10. Hu, Z., Nie, F., Chang, W., Hao, S., Wang, R., Li, X.: Multi-view spectral clustering via sparse graph learning. Neurocomputing **384**, 1–10 (2020)
11. Hu, Z., Nie, F., Wang, R., Li, X.: Multi-view spectral clustering via integrating nonnegative embedding and spectral embedding. Inform. Fusion **55**, 251–259 (2020)
12. Huang, D., Wang, C.D., Peng, H., Lai, J., Kwoh, C.K.: Enhanced ensemble clustering via fast propagation of cluster-wise similarities. IEEE Trans. Syst. Man Cybern. Syst. **51**(1), 508–520 (2021)
13. Huang, S., Kang, Z., Tsang, I.W., Xu, Z.: Auto-weighted multi-view clustering via kernelized graph learning. Pattern Recogn. **88**, 174–184 (2019)

14. Huang, S., Kang, Z., Xu, Z.: Auto-weighted multi-view clustering via deep matrix decomposition. Pattern Recogn. **97**, 107015 (2020)
15. Huang, Z., Ren, Y., Pu, X., Pan, L., Yao, D., Yu, G.: Dual self-paced multi-view clustering. Neural Netw. **140**, 184–192 (2021)
16. Kerenidis, I., Landman, J.: Quantum spectral clustering. Phys. Rev. A **103**(4), 042415 (2021)
17. Kumar, A., Daumé, H.: A co-training approach for multi-view spectral clustering. In: Proceedings of the 28th International Conference on Machine Learning, ICML'11, pp. 393–400. Madison (2011)
18. Kumar, A., Rai, P., Daume, H.: Co-regularized multi-view spectral clustering. In: Shawe-Taylor, J., Zemel, R., Bartlett, P., Pereira, F., Weinberger, K. (eds.) Advances in Neural Information Processing Systems, vol. 24. Curran Associates (2011)
19. Li, Z., Nie, F., Chang, X., Nie, L., Zhang, H., Yang, Y.: Rank-constrained spectral clustering with flexible embedding. IEEE Trans. Neural Netw. Learn. Syst. **29**(12), 6073–6082 (2018)
20. Lin, L., Tang, C., Dong, G., Chen, Z., Pan, Z., Liu, J., Yang, Y., Shi, J., Ji, R., Hong, W.: Spectral clustering to analyze the hidden events in single-molecule break junctions. J. Phys. Chem. C **125**(6), 3623–3630 (2021)
21. Liu, K., Li, X., Zhu, Z., Brand, L., Wang, H.: Factor-bounded nonnegative matrix factorization. ACM Trans. Knowl. Discovery Data (TKDD) **15**(6), 1–18 (2021)
22. Lu, H., Liu, S., Wei, H., Chen, C., Geng, X.: Deep multi-kernel auto-encoder network for clustering brain functional connectivity data. Neural Netw. **135**, 148–157 (2021)
23. Ma, J., Zhang, Y., Zhang, L.: Discriminative subspace matrix factorization for multiview data clustering. Pattern Recogn. **111**, 107676 (2021)
24. Nie, F., Wang, X., Jordan, M.I., Huang, H.: The constrained Laplacian rank algorithm for graph-based clustering. In: AAAI, pp. 1969–1976 (2016)
25. Ren, Z., Sun, Q.: Simultaneous global and local graph structure preserving for multiple kernel clustering. IEEE Trans. Neural Netw. Learn. Syst. **32**(5), 1839–1851 (2021). https://doi.org/10.1109/TNNLS.2020.2991366
26. Sellami, L., Alaya, B.: Samnet: self-adaptative multi-kernel clustering algorithm for urban vanets. Veh. Commu. **29**, 100332 (2021)
27. Sharma, K.K., Seal, A.: Multi-view spectral clustering for uncertain objects. Inform. Sci. **547**, 723–745 (2021)
28. Sun, G., Cong, Y., Dong, J., Liu, Y., Ding, Z., Yu, H.: What and how: generalized lifelong spectral clustering via dual memory. IEEE Trans. Pattern Anal. Mach. Intell. **44**(7), 3895–3908 (2021)
29. Trigeorgis, G., Bousmalis, K., Zafeiriou, S., Schuller, B.W.: A deep matrix factorization method for learning attribute representations. IEEE Trans. Pattern Anal. Mach. Intell. **39**(3), 417–429 (2016)
30. Tzortzis, G., Likas, A.: Kernel-based weighted multi-view clustering. In: 2012 IEEE 12th International Conference on Data Mining, pp. 675–684. IEEE, Piscataway (2012)
31. Von Luxburg, U.: A tutorial on spectral clustering. Stat. Comput. **17**(4), 395–416 (2007)
32. Wang, Q., He, X., Jiang, X., Li, X.: Robust bi-stochastic graph regularized matrix factorization for data clustering. IEEE Trans. Pattern Anal. Mach. Intell. **44**(1), 390–403 (2020)
33. White, M., Yu, Y., Zhang, X., Schuurmans, D.: Convex multi-view subspace learning. In: NIPS, pp. 1682–1690. Lake Tahoe (2012)
34. Wu, Z., Liu, S., Ding, C., Ren, Z., Xie, S.: Learning graph similarity with large spectral gap. IEEE Trans. Syst. Man Cybern. Syst. **51**(3), 1590–1600 (2019)
35. Xu, Y.M., Wang, C.D., Lai, J.H.: Weighted multi-view clustering with feature selection. Pattern Recogn. **53**, 25–35 (2016)
36. Yang, Z., Liang, N., Yan, W., Li, Z., Xie, S.: Uniform distribution non-negative matrix factorization for multiview clustering. IEEE Trans. Cybern. **51**(6), 3249–3262. (2020)
37. Zhan, K., Nie, F., Wang, J., Yang, Y.: Multiview consensus graph clustering. IEEE Trans. Image Process. **28**(3), 1261–1270 (2019)
38. Zhang, G.Y., Zhou, Y.R., He, X.Y., Wang, C.D., Huang, D.: One-step kernel multi-view subspace clustering. Knowl.-Based Syst. **189**, 105126 (2020)

39. Zhu, W., Nie, F., Li, X.: Fast spectral clustering with efficient large graph construction. In: 2017 IEEE International Conference on Acoustics, Speech and Signal Processing (ICASSP), pp. 2492–2496 (2017)
40. Zhu, X., Zhang, S., He, W., Hu, R., Lei, C., Zhu, P.: One-step multi-view spectral clustering. IEEE Trans. Knowl. Data Eng. **31**(10), 2022–2034 (2019). https://doi.org/10.1109/TKDE.2018.2873378
41. Zhu, X., Zhang, S., Zhu, Y., Zheng, W., Yang, Y.: Self-weighted multi-view fuzzy clustering. ACM Trans. Knowl. Discovery Data (TKDD) **14**(4), 1–17 (2020)

Chapter 7
Uber's Contribution to Faster Deep Learning: A Case Study in Distributed Model Training

Hamid Mahmoodabadi

Abstract This chapter delves into the fascinating realm of deep learning and its practical implications. It offers valuable insights that blend the scientific and technical aspects of distributed model training using the HOROVOD library in Python. This chapter's significance lies in its ability to address a crucial need within the overarching theme of "data clustering." With the explosive growth of data in today's world, efficient and scalable deep learning methods are indispensable for clustering, processing, and deriving meaningful insights from massive datasets. HOROVOD's role in enabling distributed model training not only accelerates the speed of deep learning but also opens up new horizons for data clustering, making it a pivotal tool for researchers, data scientists, and engineers seeking to harness the full potential of their data-driven endeavors.

Keywords Distributed deep learning · Distributed model training · HOROVOD · Distributed computing · Distributed training · Parallel computing · Gradient aggregation · Fault tolerance · Ring-Allreduce algorithm

7.1 Introduction to Distributed Model Training

The explosion of data volume and complexity in recent years has fueled the demand for advanced machine learning models that can unlock valuable insights from these vast resources. Deep learning, a powerful subset of machine learning, has emerged as a champion in tackling complex tasks like image recognition, natural language processing, and speech recognition. However, training these intricate models often demands immense computational power and can become prohibitively time-consuming, especially for large-scale datasets.

Enter distributed model training, a game-changer in addressing the computational hurdles associated with training deep learning on massive datasets. This

H. Mahmoodabadi (✉)
Securities and Exchange Organization of Iran, Tehran, Iran

F. Dornaika et al. (eds.), *Advances in Data Clustering*,
https://doi.org/10.1007/978-981-97-7679-5_7

approach leverages the power of parallel computing by distributing the computational workload across multiple processing units, such as CPUs or GPUs. By spreading the work across these units, distributed training frameworks enable the parallelization of the training process, significantly reducing training time and boosting overall efficiency. This allows researchers and practitioners to train larger, more complex models while harnessing the full potential of modern hardware infrastructure.

The core concept of distributed training draws inspiration from the principles of parallel computing, where tasks are divided into smaller subtasks that can be executed concurrently across multiple computing nodes. In the context of deep learning, this translates to partitioning both the training data and model parameters across distributed computing resources and orchestrating the communication and synchronization of information throughout the training process.

However, achieving efficient communication and synchronization among these distributed nodes while maintaining model accuracy and convergence presents a significant challenge. This necessitates the development of specialized algorithms and techniques for distributed optimization, parameter updates, and gradient aggregation. Additionally, distributed training frameworks must be designed to withstand potential failures, scale seamlessly, and effectively allocate resources in dynamic computing environments.

This chapter delves into the exploration of the HOROVOD, a framework specifically designed to accelerate the training of deep learning models by distributing the computational load across multiple machines [1, 2].

7.1.1 Definition of Distributed Model Training

Distributed model training, within the domain of machine learning and specifically deep learning, refers to the process of training a model across multiple computing devices or nodes simultaneously. Unlike traditional model training approaches that rely on a single computing unit to process and update model parameters, distributed model training leverages distributed computing resources to divide the computational workload and accelerate the training process.

At its core, distributed model training involves partitioning the training data and model parameters across multiple computing nodes and orchestrating the synchronization and communication of information during the training process. This parallelization of tasks enables the training of larger and more complex models, as well as the processing of massive datasets that may not fit into the memory of a single computing device.

The primary objective of distributed model training is to improve training efficiency and scalability by harnessing the computational power of multiple computing units in parallel. By distributing the computational workload, distributed training frameworks can significantly reduce training time and enable rapid experimentation and iteration required for model development.

Central to distributed model training is the implementation of specialized algorithms and techniques for distributed optimization, parameter updates, and gradient aggregation. These algorithms must be designed to ensure efficient communication and synchronization among distributed computing nodes while maintaining model accuracy and convergence. Additionally, distributed training frameworks must address challenges such as fault tolerance, scalability, and resource allocation in dynamic computing environments.

Distributed model training plays a crucial role in enabling the development of advanced machine learning models capable of addressing complex tasks and handling massive datasets. By accelerating the training process and facilitating the training of larger models, distributed model training contributes to the advancement of artificial intelligence and data-driven research.

7.1.2 Benefits of Distributing the Training Process

The burgeoning field of deep learning has revolutionized various domains, from natural language processing to robotics, by unlocking the power of complex neural networks. However, training these networks often demands substantial computational resources, especially when dealing with large-scale datasets and intricate architectures. This chapter delves into distributed model training, a paradigm that leverages distributed computing to tackle this challenge and propel deep learning advancements.

7.1.2.1 Parallelization for Speed

At its core, distributed training distributes the computational workload across multiple computing devices or nodes, enabling parallel processing. This parallelization significantly reduces training time by harnessing the collective power of multiple machines simultaneously. Frameworks like TensorFlow and PyTorch offer robust tools for orchestrating this parallelization, making it accessible to researchers and practitioners alike.

7.1.2.2 Scalability for Big Data

As datasets continue to balloon in size, distributed training emerges as a critical tool for handling these massive data troves. By partitioning the data across distributed nodes, frameworks enable efficient processing and analysis, alleviating the memory limitations of single machines. This scalability empowers researchers to tackle real-world problems involving vast datasets that were previously intractable.

7.1.2.3 Complexity Unleashed

Distributed training unlocks the potential for training more intricate and powerful models. Deep neural networks with numerous parameters often require immense computational resources, and distributing the training process across multiple nodes alleviates this bottleneck. This empowers researchers to explore sophisticated model architectures and experiment with diverse hyperparameters, ultimately leading to enhanced model performance.

7.1.2.4 Beyond Speed: Efficiency and Reliability

The benefits of distributed training extend beyond mere speed. By utilizing multiple devices in parallel, these frameworks optimize hardware utilization, minimizing idle time and boosting overall computational efficiency. Additionally, they often incorporate fault tolerance mechanisms, ensuring the training process continues uninterrupted even in the face of hardware failures or network disruptions. This resilience guarantees the successful completion of training runs, even in dynamic computing environments.

7.1.2.5 Convergence: The Key to Efficiency

Faster convergence, where the model parameters reach an optimal state, is paramount for efficient and effective training. Distributed training frameworks employ various strategies to accelerate convergence, such as:

- **Parallelized Computation:** As discussed earlier, parallel processing significantly reduces the time required for each iteration, leading to faster convergence.
- **Efficient Communication:** Specialized communication protocols minimize communication overhead and latency, ensuring swift exchange of information among distributed nodes, thereby expediting convergence.
- **Optimized Gradient Aggregation:** Efficient algorithms aggregate gradients computed across nodes while preserving accuracy, minimizing the computational overhead associated with gradient aggregation, and accelerating convergence.
- **Dynamic Resource Allocation:** Dynamically allocating resources based on computational demands optimizes resource utilization and expedites convergence by assigning more resources to compute-intensive tasks.

In conclusion, distributed model training stands as a cornerstone of modern deep learning, enabling researchers and practitioners to train larger and more complex models, process massive datasets, and achieve superior results within reasonable timeframes. By leveraging distributed computing to tackle the computational demands of deep learning, this paradigm continues to accelerate innovation and fuel the development of advanced machine learning systems capable of addressing complex real-world challenges.

7.2 The HOROVOD Library

The landscape of deep learning has witnessed unprecedented growth in recent years, driven by the demand for advanced machine learning models capable of processing vast amounts of data. Among the challenges faced in this pursuit, the need for efficient distributed model training has become increasingly paramount. In addressing this challenge, HOROVOD, an open-source distributed training framework developed by Uber Engineering, has emerged as a pivotal tool. HOROVOD is designed to expedite the training of deep neural networks by harnessing the power of distributed computing resources.

The primary motivation behind HOROVOD lies in its seamless integration with major deep learning frameworks, including TensorFlow, PyTorch, and MXNet. This integration significantly eases the adoption of distributed training, allowing researchers and practitioners to leverage the capabilities of HOROVOD without necessitating extensive modifications to their existing codebases. This interoperability is a testament to HOROVOD's commitment to enhancing accessibility and usability in the domain of distributed model training [3].

7.2.1 Features of HOROVOD

HOROVOD stands out for its innovative feature set, making it a compelling choice for researchers and practitioners alike. This section delves into the key attributes of HOROVOD, highlighting its strengths and exploring its potential to revolutionize the landscape of distributed deep learning.

- **Seamless Integration with Deep Learning Frameworks:** A hallmark of HOROVOD is its compatibility with major deep learning frameworks, such as TensorFlow and PyTorch. This compatibility stems from its ability to seamlessly integrate with existing codebases and workflows, eliminating the need for significant code modifications. This feature empowers users to leverage their existing expertise and investments, accelerating the adoption of distributed training methodologies.
- **Efficient Communication and Synchronization:** Communication and synchronization are fundamental aspects of distributed training, and HOROVOD excels in both. The framework employs the ring-allreduce algorithm, an optimization technique that minimizes communication overhead by efficiently exchanging information among distributed nodes. This translates to significant performance gains, particularly in large-scale training scenarios where communication costs can become a bottleneck.
- **Flexibility Through Diverse Training Strategies:** HOROVOD caters to a wide range of use cases by supporting various distributed training strategies. Users can choose between data parallelism, where the training data is split across nodes, and model parallelism, where the model itself is partitioned. This flexibility

empowers researchers to select the most appropriate strategy based on their specific requirements, such as the size and structure of their model and the available computational resources.

- **Dynamic Resource Allocation and Fault Tolerance:** To ensure efficient resource utilization, HOROVOD incorporates dynamic resource allocation capabilities. This feature allows the framework to adapt to varying computational demands, intelligently allocating resources where they are needed most. Additionally, HOROVOD boasts robust fault tolerance mechanisms that safeguard against hardware failures or network disruptions. These mechanisms contribute to the overall stability and reliability of the framework, ensuring the smooth execution of distributed training even in challenging environments.

In conclusion, HOROVOD stands as a testament to the power and potential of distributed training frameworks. Its seamless integration with popular deep learning frameworks, efficient communication mechanisms, support for diverse training strategies, and dynamic resource allocation capabilities make it a compelling choice for researchers and practitioners alike. As the field of deep learning continues to evolve, HOROVOD is poised to play a pivotal role in unlocking the full potential of parallel computing architectures, shaping the future of deep learning research and development.

7.2.2 Functionalities of HOROVOD

HOROVOD's functionalities are crafted to address the intricacies of distributed model training comprehensively. At its core, HOROVOD excels in parallelizing the computational workload across multiple GPUs or compute nodes. This parallelization is achieved through efficient communication and synchronization mechanisms, with a focus on minimizing communication overhead and maximizing scalability. The implementation of advanced algorithms, such as ring-allreduce, underscores HOROVOD's commitment to achieving high performance, even on large-scale distributed systems.

Additionally, HOROVOD offers support for various distributed training strategies, including data parallelism and model parallelism. This flexibility empowers users to tailor their approach based on specific hardware configurations and training objectives. The framework's ability to dynamically adjust resource allocation further enhances its functionality, allowing for optimal utilization of computational resources in diverse environments.

Fault tolerance mechanisms embedded in HOROVOD contribute to the reliability and robustness of distributed training. These mechanisms enable the framework to gracefully recover from hardware failures or network disruptions, minimizing disruptions to the training process and ensuring the continuity of model training.

7.3 Case Study: Uber's Contribution

As a key player in the technology landscape, Uber has demonstrably impacted the field of deep learning through its innovative methodologies in distributed model training. Recognizing the potential of large-scale computing, Uber actively invests in research and development, leading to the creation of novel algorithms and frameworks that significantly expedite the training process for complex neural networks. These advancements not only translate to enhanced performance within Uber's own machine learning models but also contribute to the broader research community by pushing the boundaries of distributed training techniques.

7.3.1 Specific Case Study Details

One specific case study that highlights Uber's contribution to distributed model training involves the development of a scalable and efficient framework for training deep neural networks on large-scale datasets. This framework, built upon distributed computing infrastructure, enables Uber to train complex models with millions of parameters in a fraction of the time compared to traditional training methods. By leveraging distributed model training, Uber has achieved significant improvements in model accuracy and performance in deep learning–related models [4].

7.3.2 Implementation of Distributed Model Training

Uber's implementation of distributed model training involves several key components, including specialized algorithms for distributed optimization, parameter updates, and gradient aggregation. These algorithms are designed to minimize communication overhead and maximize scalability, ensuring efficient training on distributed computing nodes. Additionally, Uber has developed tools and libraries, such as HOROVOD, to streamline the implementation of distributed model training across different deep learning frameworks.

7.3.3 Practical Example: Using HOROVOD

In this section, we provide a practical example of using the HOROVOD library with PyTorch for distributed deep learning tasks [3].

7.3.3.1 Installation

Before getting started, ensure that you have HOROVOD installed. You can install HOROVOD using pip:

Package Installation Command

```
pip install HOROVOD
```

Additionally, make sure to install the necessary dependencies for PyTorch and any other libraries you plan to use.

7.3.3.2 Setting Up HOROVOD

Once HOROVOD is installed, setting it up for PyTorch is straightforward. Here's a basic example of how to initialize HOROVOD in your PyTorch script:

Setting up HOROVOD For Pytorch

```
import torch
import HOROVOD.torch as hvd

# Initialize HOROVOD
hvd.init()

# Pin GPU to be used to process local rank (one GPU per process)
torch.cuda.set_device(hvd.local_rank())
```

This code initializes HOROVOD and sets the GPU to be used based on the local rank.

7.3.3.3 Example Usage

Now, let us demonstrate how to use HOROVOD for distributed training with a simple PyTorch script. We will use a basic example of training a neural network on the MNIST dataset.

Full Code

```
import torch
import torch.nn as nn
import torch.optim as optim
from torchvision import datasets, transforms
import HOROVOD.torch as hvd

# Initialize HOROVOD
hvd.init()

# Pin GPU to be used to process local rank (one GPU per process)
torch.cuda.set_device(hvd.local_rank())

# Define your model
class Net(nn.Module):
    def __init__(self):
        super(Net, self).__init__()
        # Define your model layers here

    def forward(self, x):
        # Define the forward pass

# Load MNIST dataset
train_dataset = datasets.MNIST('data', train=True, download=True,
                               transform=transforms.Compose([
                                   transforms.ToTensor(),
                                   transforms.Normalize((0.1307,),
                                   (0.3081,))
                               ]))

# Use DistributedSampler to distribute the dataset across nodes
train_sampler = torch.utils.data.distributed.DistributedSampler
  (train_dataset, num_replicas=hvd.size(), rank=hvd.rank())
train_loader = torch.utils.data.DataLoader(train_dataset, batch_size=64,
  sampler=train_sampler)

# Define your model, loss function, and optimizer
model = Net().cuda()
criterion = nn.CrossEntropyLoss()
optimizer = optim.SGD(model.parameters(), lr=0.01)

# Add HOROVOD DistributedOptimizer
optimizer = hvd.DistributedOptimizer(optimizer, named_parameters=
  model.named_parameters())

# Training loop
for epoch in range(epochs):
    for batch_idx, (data, target) in enumerate(train_loader):
        data, target = data.cuda(), target.cuda()
        optimizer.zero_grad()
        output = model(data)
```

```
        loss = criterion(output, target)
        loss.backward()
        optimizer.step()

# Save the model if needed
torch.save(model.state_dict(), 'model.pth')
```

This example demonstrates how to use HOROVOD to distribute the training of a PyTorch model across multiple GPUs or nodes. You can adjust the model architecture, dataset, and training parameters as needed for your specific task.

7.3.4 Scientific and Technical Aspects

From a scientific and technical perspective, Uber's contribution to distributed model training contains a range of innovative approaches and methodologies. This includes advancements in distributed optimization techniques, such as decentralized gradient aggregation and adaptive learning rate scheduling, which improve the convergence and efficiency of distributed training algorithms. Additionally, Uber's research in distributed model parallelism and data parallelism has led to novel techniques for scaling deep neural networks across distributed computing nodes while maintaining model accuracy and performance.

7.3.5 Challenges and Solutions

Despite the advancements made by Uber in distributed model training, several challenges remain, including optimizing resource utilization, managing data dependencies, and ensuring fault tolerance in distributed computing environments. To address these challenges, Uber has developed solutions such as dynamic resource allocation strategies, data partitioning techniques, and fault tolerance mechanisms, which enhance the reliability and scalability of distributed model training frameworks.

7.3.6 Results and Impact

Uber's contributions to distributed model training have had a significant impact on the field of deep learning, both within the company and across the broader research community. By accelerating the training of deep neural networks, Uber has improved the performance of its machine learning models, leading to enhanced user experiences and operational efficiencies. Furthermore, Uber's open-source contributions, such as HOROVOD, have democratized distributed model training, empowering researchers and practitioners to leverage advanced techniques for training deep learning models at scale. Overall, Uber's contribution to distributed model training has paved the way for transformative advancements in deep learning and has established the company as a leader in the development of scalable and efficient machine learning solutions.

7.4 Conclusion

This chapter explored the exciting world of distributed model training, emphasizing its importance and showcasing its capabilities through the lens of HOROVOD. We have seen how it tackles the data deluge by spreading training across multiple machines, making deep learning with massive datasets faster and more efficient. This efficiency unlocks valuable insights, propelling scientific progress and innovation in AI and machine learning. Moreover, this chapter highlights the transformative impact of distributed model training, spearheaded by frameworks like HOROVOD, on deep learning and beyond. By accelerating training, processing massive data efficiently, and fueling innovation across various domains, it holds immense potential to shape the future of AI, machine learning, and data science. As we delve deeper into its potential, we embark on a journey to unlock new knowledge frontiers, drive technological advancements, and tackle the complex challenges of our time.

References

1. Zhang, Z., Chang, C., Lin, H., Wang, Y., Arora, R., Jin, X.: Is network the bottleneck of distributed training? In: Proceedings of the Workshop on Network Meets AI & ML (NetAI '20). Association for Computing Machinery, New York (2020), pp. 8–13. https://doi.org/10.1145/3405671.3405810
2. Min, Z., Canady, R.E., Ghosh, U., Gokhale, A.S., Hakiri, A.: Tools and techniques for privacy-aware, edge-centric distributed deep learning. In: Proceedings of the Workshop on Distributed Infrastructures for Deep Learning (DIDL'20). Association for Computing Machinery, New York (2021), pp. 7–12. https://doi.org/10.1145/3429882.3430105
3. The Horovod Authors, HOROVOD Documentation, Horovod with PyTorch (2019). https://horovod.readthedocs.io/en/stable/
4. Panda, D.K., Awan, A.A., Subramoni, H.: High performance distributed deep learning: a beginner's guide. In: Proceedings of the 24th Symposium on Principles and Practice of Parallel Programming (PPoPP '19). Association for Computing Machinery, New York (2019), pp. 452–454. https://doi.org/10.1145/3293883.3302260

Chapter 8
Auto-weighted Multi-view Clustering with Unified Binary Representation and Deep Initialization

Khamis Houfar, Fadi Dornaika, Djamel Samai, Azeddine Benlamoudi, Khaled Bensid, and Abdelmalik Taleb-Ahmed

Abstract Clustering, as an integral part of exploratory data analysis, has gained renewed attention due to the prevalence of multi-representational or multi-view real-world data. However, its application becomes increasingly challenging in the face of large and heterogeneous datasets. Notably, existing techniques aimed at enhancing computational efficiency often possess drawbacks, such as assigning equal or static weights to views and samples, limiting the utilization of common and complementary features. Additionally, many methods execute the clustering task with arbitrary initialization, neglecting the rich structure of the joint discrete representation. In response to these challenges, this chapter introduces a novel approach named "auto-weighted binary multi-view clustering via deep initialization" designed for large-scale multi-view clustering. Two primary scenarios guide our approach: first, the differentiation between views based on sample importance, utilizing a dynamic learning strategy for automatic weighting of views and samples. Second, in the context of initializing binary clustering, we leverage a new CNN feature and employ a low-dimensional binary embedding, capitalizing on the efficient capabilities of Fourier mapping. Our proposed approach simultaneously learns a joint discrete representation and conducts direct clustering through constrained binary

K. Houfar · D. Samai · A. Benlamoudi · K. Bensid
University of Ouargla, Faculté des Nouvelles Technologies de l'information et de la Communication, Laboratoire de Génie Électrique (LAGE), Ouargla, Algeria
e-mail: samai.djamel@univ-ouargla.dz; benlamoudi.azeddine@univ-ouargla.dz; bensid.khaled@univ-ouargla.dz

F. Dornaika (✉)
University of the Basque Country UPV/EHUm, San Sebastian, Spain

IKERBASQUE, Basque Foundation for Science, Bilbao, Spain
e-mail: fadi.dornaika@whu.eus; fadi.dornaika@ehu.eus

A. Taleb-Ahmed
Institut d'Electronique de Microélectronique et de Nanotechnologie (IEMN), UMR 8520, Université Polytechnique Hauts de France, Université de Lille, CNRS, Valenciennes, France
e-mail: Abdelmalik.Taleb-Ahmed@uphf.fr

F. Dornaika et al. (eds.), *Advances in Data Clustering*,
https://doi.org/10.1007/978-981-97-7679-5_8

matrix factorization, solving the optimization problem in a unified learning model. Experimental results on several challenging datasets showcase the effectiveness and superiority of our approach over state-of-the-art methods, measured in terms of accuracy, normalized mutual information, and purity.

Keywords Multi-view clustering · Large scale datasets · Anchors · Discrete representation · Bidirectional FFT

8.1 Introduction

In data mining, machine learning, and image processing applications, data is often represented by multiple feature sets, known as multi-view data. Each view corresponds to a distinct visual descriptor, such as HOG, SIFT, GIST, or LBP, providing both common and complementary information essential for successful multi-view learning [8]. Within the realm of multi-view learning, this work focuses on unsupervised clustering techniques.

Traditional clustering algorithms designed for single-view data may not effectively handle multi-view data, where concatenating all views and applying advanced clustering algorithms may lead to overfitting due to information redundancy. Multi-view clustering (MVC) methods [6, 21] can be broadly categorized into three main classes: (1) common feature subspace, involving techniques like canonical correlation analysis (CCA) to minimize cross-correlation error and subsequent data grouping; (2) multi-view spectral clustering, which constructs multiple graphs to characterize geometric structure and performs data partitioning; (3) multi-view nonnegative matrix factorization (NMF) clustering, which involves matrix factorization for data partitioning.

Hashing techniques, also known as binary code learning [4, 5, 24], have gained importance in big data analysis, enabling fast Hamming distance computation and reduced memory requirements. Multi-view hashing approaches have emerged for visual search and rapid object detection [24, 28]. These methods embed high-dimensional feature vectors into low-dimensional binary codes through projections, facilitating information exchange between multiple views while preserving the intrinsic aspects of the original space.

Despite significant progress, existing multi-view learning algorithms often exhibit three main drawbacks:

1. The majority of existing models adopt equal or fixed weights, introducing additional parameters to estimate the contribution of each view, leading to suboptimal representation learning.
2. Many current models treat all samples equally during the clustering process.
3. Proposed methods often lack a viable and informative initialization for the binary codes in the clustering task, resulting in suboptimal local optima.

To address the challenges at hand, this chapter introduces a novel method for multi-view clustering: auto-weighted binary multi-view clustering (AW-BMVC) via deep initialization. Our motivation encompasses two key aspects: data discovery and analysis model. In the realm of data discovery, our goal is to harness the rich attributes found in real-world visual applications, spanning different views and modalities, and integrate them into a cohesive binary representation. This integration seeks to enhance feature interpretability and linear separability by exploring nonlinear structures facilitated by kernel advantages. In the analysis model domain, we focus on three pivotal steps: assessing view diversity, accounting for sample variance, and refining clustering initialization. The algorithm autonomously gauges the importance of each view and determines sample weights based on learning loss—implicitly for views and explicitly for samples. The automatic weighting strategy aims to mitigate the impact of noisy or outlier views/samples, leading to a more robust estimation without the need for additional manually adjusted parameters. Similar successful approaches employing automatic and adaptive weighting have been observed in recent machine learning algorithms.

The binary embedding of samples involves mapping from the kernelized higher-dimensional real space of features to the lower-dimensional Hamming space, resulting in common binary codes. This binary representation offers a dual advantage: mitigating noise affecting real-valued features in different views and enhancing optimization efficiency by simplifying certain steps in the process.

The final crucial step involves developing an efficient strategy to guide the clustering model toward the optimal point. Notably, for many multi-view clustering approaches, the primary challenge lies in achieving effective fusion, aiming to approximate multi-view data in a unified representation while fully exploiting the diverse information inherent in multiple views.

The following are the main contributions:

1. To leverage the diversity inherent in data with multiple views, we employ an automatically weighted strategy that governs the pairwise importance of each sample and each view independently. View weights are implicitly derived from the square root of the view objective function, while sample weights are explicitly estimated.
2. Furthermore, we introduce an objective function that facilitates the joint estimation of several critical entities, including the common binary code of the data, two sets of weights, the view-based mapping from the nonlinear representation to the common binary code space, binary centroids, and the cluster assignment matrix.
3. To kickstart the optimization process effectively, we extract deep features from the Vgg16 network, obtaining a robust initialization for our proposed method. These features are then mapped to a low-dimensional Hamming space using a bidirectional fast Fourier transform technique (BD-FFT). The resulting binary vectors serve as an initialization for our iterative clustering algorithm.
4. By harnessing the presented objective function and alternating optimization scheme, our proposed method demonstrates superior performance compared

to numerous state-of-the-art multi-view clustering techniques, including those based on real values.

The subsequent sections of the chapter unfold as follows: Sect. 8.2 delves into key concepts and related work. Section 8.3 comprehensively elucidates the proposed methodology. Extensive performance analysis is presented in Sect. 8.4. Finally, Sect. 8.5 encapsulates the chapter with concluding remarks and outlines avenues for future research.

8.2 Related Work

Multi-view clustering (MVC) stands as a captivating subject in machine learning, prompting exploration through diverse methodologies.

Prior to delving into related work, an overview of MVC methods is presented. A significant portion of multi-view approaches can be categorized into distinct classes, including spectral clustering [32], graph-based clustering [7], and subspace clustering [14].

For a succinct overview, here are brief descriptions of noteworthy works:

- **Robust multi-view spectral clustering (RMSC)** [15]: This approach initiates by constructing a graph for each view. Subsequently, it utilizes a joint transition probability matrix, formulated through low-rank constraints and sparse decomposition. The resulting matrix serves as a key input for standard Markov chain clustering. However, it has limitations for large datasets and fails to accommodate the flexible structure of local manifolds, impacting the agreement between views.
- **Diversity-induced multi-view subspace clustering (DiMSC)** [25]: DiMSC is a self-representation based subspace clustering method. It directly represents each data point with the data collection in the original view. It explores diversity between multiple views using the Hilbert-Schmidt independence criterion (HSIC) to estimate differences across different representations. The partitioning result is obtained through a subsequent spectral clustering strategy. DiMSC prioritizes merging information over enhancing the ability to represent features.
- **Adaptively weighted procrustes (AWP)** [3]: AWP introduces a spectral embedding of kernels from different views, alongside the Procrustes analysis technique, to learn a unified cluster indicator matrix that accommodates all spectral embeddings. While it has a lower computational cost compared to other graph-based methods, it necessitates a post-processing step involving spectral rotation to obtain cluster labels.
- **Weighted multi-view spectral clustering (WMSC)** [32]: This method involves generating a normalized Laplacian matrix for each view. Subsequently, it constructs a combined joint Laplacian matrix by considering learned weights, thereby discerning the contribution of each respective view. The approach utilizes spectral clustering to predict labels, taking into account the principle of spectral perturbation. This principle aims to minimize the clustering discrepancy between

each selected view and the joint clustering. The method is rooted in measuring the proximity of subspaces spanned by eigenvectors using canonical angles to capture differences in cluster ability.

- **One-step multi-view spectral clustering (OMSC)** [26]: This method is designed to simultaneously learn the affinity matrix for each view and the consensus graph in the intrinsic space. It involves assigning a projection matrix for each view to map the constructed affinities into a low-dimensional space. Additionally, dynamic view weighting is incorporated to quantify the importance of each view. Rather than applying a clustering algorithm, the method generates an implicit partitioning result by permuting the consensus affinity matrix. This permutation ensures the grouping of the data into connected components based on the Laplacian rule and Ky Fan's theorem. It is worth noting that this method exhibits sensitivity to hyperparameters.
- **Large-scale multi-view subspace clustering (LMVSC)** [30]: This method employs multiple anchor graphs to reduce time complexity. It involves computing a double stochastic similarity matrix and performing eigenvalue decomposition on this small graph. The final clustering is achieved by applying K-means to the embedded space. However, it should be noted that this approach does not thoroughly explore the nonlinear high-order correlation between the consensus latent subspace and the different view spaces.
- **Nonnegative embedding and spectral embedding (NESE)** [29]: This approach constructs graphs from different views and considers spectral embedding. It combines view-based graph matrices and spectral representation simultaneously, inspired by a learning model of symmetric matrix factorization. The goal is to iteratively estimate a consistent nonnegative embedding that directly reveals a joint partitioning result. Unlike AWP, this method estimates clustering labels directly without post-processing. However, it may be sensitive to noise and outliers as complementary information is merged into a nonnegative embedding matrix.
- **Graph-based multi-view clustering (GMC)** [7]: This method combines graph construction, graph fusion, and data clustering into a unified framework. It learns the graph of each view and the unified graph of all views by mutual reinforcement. The unified graph, subject to a rank constraint, is then used to directly partition the data points into clusters. Notably, this method automatically assigns weights to each graph matrix to obtain a unified graph matrix.
- **Smoothed multi-view subspace clustering (SMVSC)** [14]: This method introduces a graph-filtering technique to obtain a smooth representation. Initially, a graph is created for each view using the probabilistic neighborhood method. Subsequently, a graph filter is applied to these graphs, and representative anchors are selected. The method involves concatenating different filtered graphs to create a joint anchor graph fusion. Finally, an eigenvalue decomposition of this matrix is performed, and clustering is carried out using K-means. SMVSC provides an alternative approach to LMVSC, utilizing graph filtering to achieve a smooth representation in each view.

- **Collaborative feature-weighted multi-view fuzzy c-means clustering (Co-FW-MVFCM)** [13]: This method introduces a feature- and view-weighted scheme by integrating two steps: local and collaborative learning. In the local step, the partitioning of each view is targeted, while the collaborative step focuses on sharing information about the membership of multiple views. Finally, global clustering is achieved by aggregating the weighted partition matrices from different views. However, a potential limitation of this model is the absence of clear criteria for selecting the optimal exponent parameter to control the view weights.

However, there are limited studies that have focused on clustering large binary data. Gong et al. [27] developed a method for binary clustering in a single view, involving two distinct steps: binary code generation and binary K-means clustering. The main drawback lies in the fact that the binary code is generated using a data-independent method, iterative quantization (ITQ). In another work [5], a two-level clustering approach was employed, breaking the link between binary representation and data partitioning. To expedite large-scale clustering of individual views, Shen et al. [17] combined binary structural SVM and conventional K-means in an optimization algorithm. However, neither method is directly applicable to large-scale multi-view clustering (MVC), and the characteristics of multi-view data have not been thoroughly explored. Additionally, the binary codes generated by Shen et al. [17] yielded unsatisfactory results due to the lack of a complete joint representation.

In contrast, Zhang et al. [31] proposed an intriguing approach called binary multi-view clustering (BMVC) to address a critical challenge in multi-view clustering, aiming for reduced computation time and storage costs.

BMVC uncovers two essential elements: collaborative discrete representation learning and binary clustering structure learning in a unified model. By considering only complementary features, this framework encodes multi-view features into a shared compact binary code. The model introduces a nonnegative normalized vector to weigh the views, with an additional adjustable parameter to balance the importance of different views.

The BMVC method encountered difficulties in effectively distinguishing between shared and individual information, potentially resulting in the loss of local structure preservation during binary code learning. In response to this challenge, the highly economized multi-binary (HSIC) method was introduced, aiming to jointly learn a common binary representation and robust discrete cluster structures [28]. The former decomposes each projection into a combination of shareable and individual projections across multiple views to capture underlying correlations, while the latter enhances computational efficiency and robustness in clustering. However, this approach is sensitive to the initialization of the binary clustering process, and its performance degrades when attempting to eliminate extra parameters and automatically learn the weighting factor for each view.

To address these limitations, our work draws inspiration from the BMVC framework. Positioned within the domain of multi-view nonnegative matrix factorization (NMF) clustering, our approach characterizes the relationship between views based on samples, employing strategies for automatic weighting of both samples and views. Clustering is conducted through a joint binary matrix factorization with a bit balance constraint [5], a common requirement in binary code learning. The initialization of the discrete representation plays a pivotal role in guiding the iterative binary clustering optimization toward an optimal solution. In alignment with this initialization concept, we propose an efficient solution that incorporates new deep features from Vgg16. These features are subsequently encoded into a set of compact binary codes using the bidirectional FFT technique [9].

However, the above work is very sensitive to the initialization of the binary clustering process, and even the performance degrades when trying to get rid of the extra parameter and learn the weighting factor of each view automatically.

8.3 The Proposed Approach

In this chapter, matrices are denoted by bold uppercase letters, while vectors are represented by bold lowercase letters. A comprehensive summary of the notations employed is provided in Table 8.1.

This section offers an in-depth explanation of our novel multi-view clustering approach, named auto-weighted binary multi-view clustering (AW-BMVC) via deep initialization. The method encompasses two central learning objectives: the creation of a common discrete representation driven by the auto-weighted sampling strategy and the auto-weighted view strategy. Simultaneously, the global objective function is initialized with a well-crafted binary matrix representation. In essence, the framework's diagram is illustrated in Fig. 8.1.

8.3.1 Anchor-Based Representation

Assume a radial basis function (RBF) that effectively consolidates various views into a single tensor with fixed dimensionality and thoroughly explores the high-order latent structure within multiple views by projecting them into a higher-dimensional space.

Consider a multi-view dataset comprising V representations (i.e., V views) for n instances, represented by a set of matrices $\{\mathbf{X}^1, \ldots., \mathbf{X}^V\}$, where $\mathbf{X}^v \in \mathbb{R}^{d_v \times n}$ is the data matrix of the vth view, and d_v is the dimensionality of data features from the vth view. It is assumed that data samples in each view are zero centered, i.e., $\sum_s \mathbf{x}_s^v = 0$, to maintain data balance. The initial step involves encoding the data using a nonlinear RBF mapping, given by the following transformation:

$$\boldsymbol{\Phi}(\mathbf{x}_s^v) = \left[\exp\left(-\frac{||\mathbf{x}_s^v - \mathbf{a}_1^v||^2}{\sigma^v}\right), \ldots, \exp\left(-\frac{||\mathbf{x}_s^v - \mathbf{a}_m^v||^2}{\sigma^v}\right)\right]^T \quad (8.1)$$

Table 8.1 Summary of the main notations

Notation	Description	Notation	Description
n	Number of samples	$(\cdot)^T$	Transpose operator
c	Number of clusters	$\mathbf{I}$	Identity matrix
V	Number of views	$h(\cdot)$	Discrete hash function
m	Number of anchors	l	Binary code length
d_v	Data dimensionality for view v	$\mathbf{B} = [\mathbf{b}_1, \ldots, \mathbf{b}_n] \in \{-1, +1\}^{l \times n}$	The common binary codes of the n samples
$\mathbf{X}^1, \ldots, \mathbf{X}^M$ where $\mathbf{X}^v \in \mathbb{R}^{d_v \times n}$	A set of V data matrices	$\mathbf{U}^v \in \mathbb{R}^{l \times m}$	The mapping matrix for the v-th view
$\mathbf{x}_s^v$	s-th sample from the vth view	α	View-weighting vector
$\mathbf{a}_1^v, \mathbf{a}_2^v, \ldots, \mathbf{a}_m^v$	A set of selected anchors from the vth view	$\mathbf{W} \in \mathbb{R}^{n \times n}$	Sample-weighting matrix (a diagonal matrix)
$\mathbf{\Phi}^v \in \mathbb{R}^{m \times n}$	Nonlinear Radial Basis Function mapping for view v	$\beta, \gamma, \lambda, \rho$	Regularization parameters
σ	Kernel width	$\mathbf{C} \in \{-1, +1\}^{l \times c}$	Clustering binary centroids
$\text{sgn}(\cdot)$	Signum operator	$\mathbf{G} \in \{0, 1\}^{c \times n}$	Clustering assignment
$\|\|\cdot\|\|_F$	Frobenius norm	$\mathbf{1}$	Column vector of ones
$Tr(\cdot)$	Trace of a matrix		

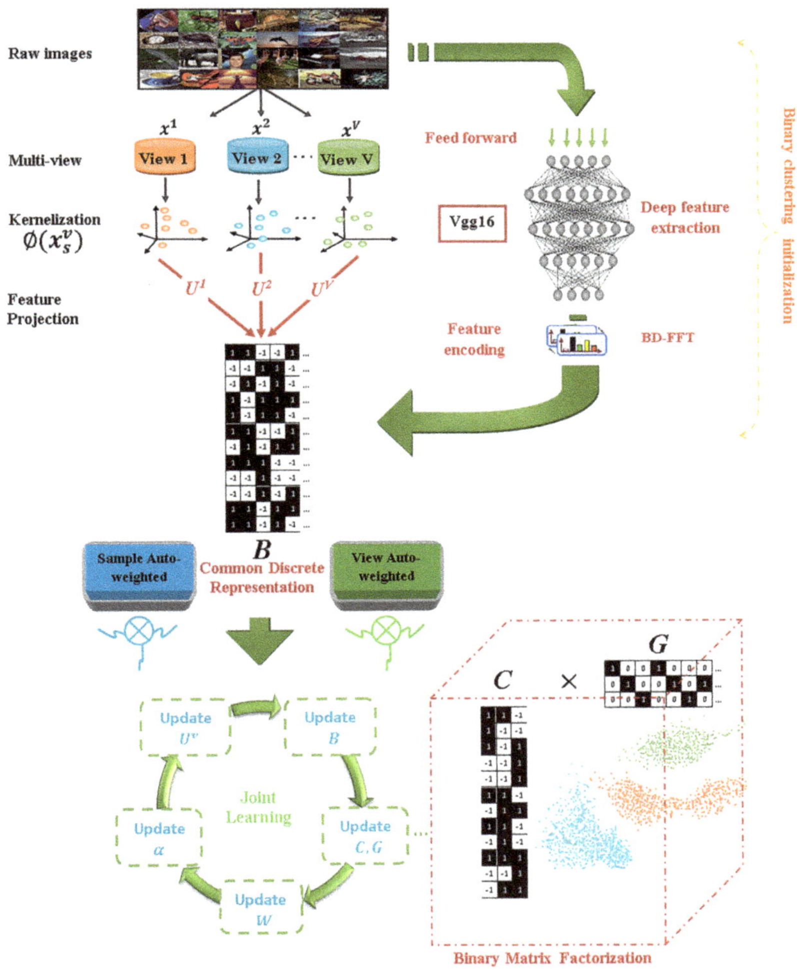

Fig. 8.1 The flowchart of the proposed method. Common discrete representation, binary clustering initialization, sample and view auto-weighting, and binary matrix factorization are integrated into a unified learning framework

In the above expression, σ^v denotes the kernel width for the vth view, $\mathbf{\Phi}(\mathbf{x}_s^v) \in \mathbb{R}^m$ represents the m-dimensional nonlinear embedding for the s-th sample from the vth view, and $\{\mathbf{a}_1^v, \mathbf{a}_2^v, \ldots, \mathbf{a}_m^v\}$ is a set of m selected anchors from the vth view. Anchors can be conceptualized as statistically representative elements of the broader dataset, obtained using the K-medoids technique for its robustness to noise [1] as opposed to random sampling or K-means.

Remark We set the number of selected anchors for each view to $m = 1000$, based on experiments outlined in [31]. It is essential to note that the kernel width parameter σ^v plays a crucial role, determining the degree of smoothing [16] and often requires careful manual tuning. Empirically, a universal adaptive scaling is established, where the global width for each view is set to the average of all the Euclidean distances between the samples and their corresponding anchors.

8.3.2 Common Discrete Representation

The primary objective of our unsupervised method is to directly perform clustering in a significantly lower-dimensional Hamming space using common binary codes, effectively compressing multiple views. To achieve this, we employ hashing as a prevalent technique for computationally efficient similarity preservation.

We introduce a discriminative hashing function to be learned for each view, aiming to quantize each $\mathbf{\Phi}(\mathbf{x}_s^v)$ into a discrete representation as follows:

$$\min_{\mathbf{U}^v, \mathbf{b}_s} \sum_{v=1}^{V} \sum_{s=1}^{n} ||\mathbf{b}_s - \mathbf{U}^v \mathbf{\Phi}(\mathbf{x}_s^v)||^2 = \min_{\mathbf{U}^v, \mathbf{B}} \sum_{v=1}^{V} ||\mathbf{B} - \mathbf{U}^v \mathbf{\Phi}(\mathbf{X}^v)||_F^2 \tag{8.2}$$

$$\mathbf{b}_s = h_s^v(\mathbf{\Phi}(\mathbf{x}_s^v); \mathbf{U}^v) = \operatorname{sgn}(\mathbf{U}^v \mathbf{\Phi}(\mathbf{x}_s^v)) \tag{8.3}$$

In the equation above, $\mathbf{B} = [\mathbf{b}_1, \ldots, \mathbf{b}_n]$ represents the common binary codes from different views (i.e., $\mathbf{x}_s^v, \forall v = 1, \ldots, V$), and $\mathbf{\Phi}(\mathbf{X}^v)$ is the matrix containing nonlinear representations of all samples in view v. It can be expressed as $\mathbf{\Phi}(\mathbf{X}^v) = [\mathbf{\Phi}(\mathbf{x}_1^v), \ldots, \mathbf{\Phi}(\mathbf{x}_n^v)]$. The mapping matrix is denoted by $\mathbf{U}^v$, and $\operatorname{sgn}(\cdot)$ stands for the element-wise sign operator. It is important to note that despite the linearity of the model in Eq. (8.2), the overall mapping from data space to the common binary code space is nonlinear due to the utilization of the nonlinear mapping $\mathbf{\Phi}(\mathbf{X}^v)$.

8.3.3 Sample View Auto-weighting

Acknowledging that different views represent the same subject from various measurements, the projections $\{\mathbf{U}^v\}_{v=1}^V$ are designed to capture consensus information maximizing similarities between different views and discerning individual characteristics' disparities. To characterize the relationship between views, implicit automatic view weighting will be adopted. Simultaneously, explicit sample weighting coefficients will be estimated in global optimization. This strategy allows for interchangeably emphasizing vital samples and promoting complementary information between different views, resulting in a comprehensive common discrete representation.

To delve deeper into the information-theoretic perspective, it is imperative to maximize the information content carried by each bit of the binary codes [23]. Embracing this notion, an additional regularizer is introduced for the binary codes

B using the maximum entropy principle [5]. Therefore, the objective is to maximize the variance of the matrix **B** given by:

$$\begin{aligned} \text{var}[\mathbf{B}] &= \frac{1}{n}\sum_{v=1}^{V} \text{var}[\mathbf{U}^v \boldsymbol{\Phi}(\mathbf{X}^v)] = \frac{1}{n}\sum_{v=1}^{V} \|\mathbf{U}^v \boldsymbol{\Phi}(\mathbf{X}^v)\|^2 \\ &= \frac{1}{n}\sum_{v=1}^{V} tr((\mathbf{U}^v \boldsymbol{\Phi}(\mathbf{X}^v))(\mathbf{U}^v \boldsymbol{\Phi}(\mathbf{X}^v))^T) \end{aligned} \tag{8.4}$$

This supplementary regularization on **B** serves to ensure a balanced partition and diminish the redundancy of the binary codes [28]. We cast the relaxed regularization as a common discrete representation learning problem:

$$\begin{aligned} \min F(\mathbf{U}^v, \mathbf{B}, \mathbf{W}) = \sum_{v=1}^{V} \Big(&||(\mathbf{B} - \mathbf{U}^v \boldsymbol{\Phi}(\mathbf{X}_s^v))\, \mathbf{W}||_F^2 + \beta\, ||\mathbf{U}^v||_F^2 \\ &- \frac{\gamma}{n} tr((\mathbf{U}^v \boldsymbol{\Phi}(\mathbf{X}^v))(\mathbf{U}^v \boldsymbol{\Phi}(\mathbf{X}^v))^T)\Big) \\ &s.t. \mathbf{B} \in \{-1, 1\}^{l \times n}, \sum_s w_s = 1, w_s > 0, \end{aligned} \tag{8.5}$$

where β and γ are two regularization parameters. The second term serves as a regularizer controlling the parameter scales, contributing to a stable solution. Here, $\mathbf{W} = \text{diag}(w_1, w_2, \ldots, w_n)$ represents the diagonal sample-weighting matrix. By learning the weights for samples, more importance is assigned to those with larger weights.

Taking inspiration from recent advancements in auto-weighted techniques [22], we present a novel formulation that eliminates the need for explicit view weight factors.

In this regard, the previous objective function is replaced with a new one, where the square root of the term to be minimized is considered. This reformulation transforms the problem into:

$$\begin{aligned} \min_{\mathbf{U}^v, \mathbf{B}, \mathbf{W}} &= \sum_{v=1}^{V} \sqrt{||(\mathbf{B} - \mathbf{U}^v \boldsymbol{\Phi}(\mathbf{X}^v) \mathbf{W}||_F^2 + \beta\, ||\mathbf{U}^v||_F^2 - \frac{\gamma}{n} tr(\mathbf{U}^v \boldsymbol{\Phi}(\mathbf{X}^v))(\mathbf{U}^v \boldsymbol{\Phi}(\mathbf{X}^v))^T)} \\ &s.t.\ \mathbf{B} \in \{-1, 1\}^{l \times n}, \sum_s w_s = 1, w_s > 0 \end{aligned} \tag{8.6}$$

Like many multi-view algorithms, this criterion implicitly assigns a weight to each view. Therefore, minimizing Eq. (8.6) is equivalent to minimizing the following:

$$\begin{aligned}\min_{\mathbf{U}^v,\mathbf{B},\mathbf{W}} = \sum_{v=1}^{V} \alpha^v \Big(&||(\mathbf{B} - \mathbf{U}^v \mathbf{\Phi}(\mathbf{X}^v))\mathbf{W}||_F^2 + \beta ||\mathbf{U}^v||_F^2 \\ &-\frac{\gamma}{n} tr(\mathbf{U}^v \mathbf{\Phi}(\mathbf{X}^v))(\mathbf{U}^v \mathbf{\Phi}(\mathbf{X}^v))^T \Big) \\ s.t. \mathbf{B} \in \{-1, 1\}^{l \times n}, &\sum_s w_s = 1, w_s > 0, \end{aligned} \tag{8.7}$$

where the auto-weight α^v is given by the following expression:

$$\alpha^v = \frac{1}{2\sqrt{||(\mathbf{B} - \mathbf{U}^v \mathbf{\Phi}(\mathbf{X}^v))\, \mathbf{W}||_F^2 + \beta ||\mathbf{U}^v||_F^2 - \frac{\gamma}{n} tr(\mathbf{U}^v \mathbf{\Phi}(\mathbf{X}^v))(\mathbf{U}^v \mathbf{\Phi}(\mathbf{X}^v))^T}} \tag{8.8}$$

8.3.4 Binary Matrix Factorization and Overall Objective Function

AW-BMVC involves the direct factorization of the learned discrete representation **B** into two matrices: the binary clustering centroids **C** and the discrete clustering indicators **G**, subject to specific constraints, using:

$$\begin{aligned}&\min_{\mathbf{C},\mathbf{g}_s} ||\mathbf{b}_s - \mathbf{C}\,\mathbf{g}_s||_F^2 \\ s.t.\ \ &\mathbf{C}^T\mathbf{1} = 0, \mathbf{C} \in \{-1, 1\}^{l \times c}, \mathbf{g}_s \in \{0, 1\}^c, \sum_i^c g_{is} = 1\end{aligned} \tag{8.9}$$

where **C** and $\mathbf{g}_s$ are the clustering centroids and the assignment vector for the sample s, respectively. The clustering centers constraint ($\mathbf{C}^T\mathbf{1} = 0$) enforces the balance condition to maximize the information of each bit. Expanding Eq. (8.9) for all samples, we obtain the following factorization problem:

$$\begin{aligned}&\min_{\mathbf{C},\mathbf{G}} ||(\mathbf{B} - \mathbf{C}\,\mathbf{G})\,\mathbf{W}||_F^2 \\ s.t.\ \ &\mathbf{C}^T\mathbf{1} = 0, \mathbf{C} \in \{-1, 1\}^{l \times c}, \mathbf{G} \in \{0, 1\}^{c \times n}, \sum_{i=1}^c G_{is} = 1\end{aligned} \tag{8.10}$$

So, the overall joint AW-BMVC is formulated as:

$$\min F(\mathbf{U}^v, \mathbf{B}, \mathbf{C}, \mathbf{G}, \mathbf{W}, \alpha) = \sum_{v=1}^{V} \alpha^v (||(\mathbf{B} - \mathbf{U}^v \boldsymbol{\Phi}(\mathbf{X}^v))\, \mathbf{W}||_F^2 +$$

$$\beta\, ||\mathbf{U}^v||_F^2 - \frac{\gamma}{n} tr((\mathbf{U}^v \boldsymbol{\Phi}(\mathbf{X}_s^v))(\mathbf{U}^v \boldsymbol{\Phi}(\mathbf{X}^v))^T) + \lambda\, ||(\mathbf{B} - \mathbf{C}\,\mathbf{G})\mathbf{W}||_F^2$$

$$s.t.\ \ \mathbf{C}^T \mathbf{1} = 0, \sum_s w_s = 1, w_s > 0, \tag{8.11}$$

$$\mathbf{B} \in \{-1, 1\}^{l \times n}, \mathbf{C} \in \{-1, 1\}^{l \times c}, \mathbf{G} \in \{0, 1\}^{c \times n}, \sum_{i=1}^{c} G_{is} = 1,$$

where λ is the regularization parameter.

It is essential to highlight the inclusion of the sample auto-weighted matrix $\mathbf{W}$ in the binary clustering learning phase. This incorporation is instrumental in preserving information and maintaining the equilibrium between the discrete representation and the binary clustering learning.

8.3.5 *Optimization*

The resolution of problem (8.11) is inherently challenging due to its combinatorial optimization nature, involving discrete constraints and the nonlinearity of the objective function. Consequently, we employ an alternating optimization scheme to break down the problem into smaller subproblems, updating it iteratively with respect to one variable while keeping the other variables fixed.

We define each step to iteratively update the mapping matrix $\mathbf{U}^v$, the discrete representation $\mathbf{B}$, the binary cluster centroids $\mathbf{C}$ and the indicator $\mathbf{G}$, the sample auto-weighting $\mathbf{W}$, and the view auto-weighting α^v.

- **Step 1: Update $\mathbf{U}^v$**, $v = 1, \ldots, V$.
  By fixing other variables, the optimization formula for $\mathbf{U}^v$ is:

$$\min F(\mathbf{U}^v) = ||(\mathbf{B} - \mathbf{U}^v \boldsymbol{\Phi}(\mathbf{X}^v))\, \mathbf{W}||_F^2 + \beta\, ||\mathbf{U}^v||_F^2 - \frac{\gamma}{n} tr((\mathbf{U}^v \boldsymbol{\Phi}(\mathbf{X}^v))(\mathbf{U}^v \boldsymbol{\Phi}(\mathbf{X}^v))^T \tag{8.12}$$

  By computing the derivative of the objective function with respective to $\mathbf{U}^v$, and setting it to 0, we can obtain the following solution:

$$\mathbf{U}^v = \mathbf{B}\,\mathbf{W}\,\mathbf{W}\,\boldsymbol{\Phi}(\mathbf{X}^v)^T \cdot \mathbf{Q} \tag{8.13}$$

  where $\mathbf{Q} = \left[\boldsymbol{\Phi}(\mathbf{X}^v)\mathbf{W}\mathbf{W}\boldsymbol{\Phi}(\mathbf{X}^v)^T - \frac{\gamma}{n}\boldsymbol{\Phi}(\mathbf{X}^v)\boldsymbol{\Phi}(\mathbf{X}^v)^T + \beta\,\mathbf{I}\right]^{-1}$.

- **Step 2: Update B**.
  The optimization formula for **B** is:

$$
\begin{aligned}
\min_{\mathbf{B}} = & \sum_{v=1}^{V} \alpha^{v} \left(||(\mathbf{B} - \mathbf{U}^{v} \boldsymbol{\Phi}(\mathbf{X}^{v}))\mathbf{W}||_F^2 \right) + \lambda \, ||(\mathbf{B} - \mathbf{CG})\mathbf{W}||_F^2 \\
= & \sum_{v=1}^{V} \alpha^{v} \, tr \left((\mathbf{BW} - \mathbf{U}^{v} \boldsymbol{\Phi}(\mathbf{X}^{v}) \, \mathbf{W})^T (\mathbf{B\,W} - \mathbf{U}^{v} \boldsymbol{\Phi}(\mathbf{X}^{v}) \, \mathbf{W}) \right) + \\
& \lambda \, tr \left((\mathbf{B\,W} - \mathbf{C\,G\,W})^T (\mathbf{B\,W} - \mathbf{C\,G\,W}) \right) \\
& = tr \left[\mathbf{B}^T \left(\sum_{v=1}^{V} \alpha^{v} \mathbf{WW}^T + \lambda \, \mathbf{WW}^T \right) \mathbf{B} \right] - 2 \\
& tr \left[\mathbf{B}^T \left(\sum_{v=1}^{V} \alpha^{v} \mathbf{U}^{v} \boldsymbol{\Phi}(\mathbf{X}^{v}) \mathbf{WW} + \lambda \, \mathbf{C\,G\,W\,W} \right) \right] + \text{cons} \\
& s.t. \;\; \mathbf{B} \in \{-1, 1\},
\end{aligned}
\tag{8.14}
$$

  The solution for **B** is given by:

$$
\mathbf{B} = \text{sgn} \left(\sum_{v=1}^{V} \alpha^{v} \mathbf{U}^{v} \boldsymbol{\Phi}(\mathbf{X}^{v}) \mathbf{W\,W} + \lambda \, \mathbf{C\,G\,W\,W} \right) \tag{8.15}
$$

- **Step 3: Update C and G**.
  The regularized optimization formula for **C** and **G** taking into account the discrete constraints will be given by:

$$
\begin{aligned}
& \min F(\mathbf{C}, \mathbf{G}) = ||(\mathbf{B} - \mathbf{CG})\mathbf{W}||_F^2 + \rho ||\mathbf{C}^T \mathbf{1}||^2 \\
& s.t. \mathbf{C} \in \{-1, 1\}^{l \times c}, \mathbf{G} \in \{0, 1\}^{c \times n}, \sum_{i} g_{is} = 1,
\end{aligned}
\tag{8.16}
$$

  We iteratively optimize the cluster centroids using the adaptive discrete proximal linearized minimization (ADPLM) technique [4]. This approach ensures that the discrete constraints are preserved throughout the optimization process.

  Update C.

  With **G** fixed, we have the following minimization problem:

$$
\min F(\mathbf{C}) = -2tr \left[(\mathbf{BW})^T (\mathbf{CGW}) \right] + \rho ||\mathbf{C}^T \mathbf{1}||^2 + \text{cons} \tag{8.17}
$$

The derivative of the obtained functional with respect to **C** is given as follows:

$$\nabla F(\mathbf{C}) = -2\mathbf{B}\,\mathbf{W}\,(\mathbf{G}\,\mathbf{W})^T + 2\rho\,\mathbf{E}\,\mathbf{C} \qquad s.t. \mathbf{C} \in \{-1, 1\}^{l\times c}, \tag{8.18}$$

where $\nabla F(\mathbf{C})$ is the gradient of $F(\mathbf{C})$ and **E** is $l \times l$ square matrix of ones.

Based on the rule of ADPLM, we update **C** in the $p+1$-th iteration by

$$\mathbf{C}^{p+1} = \text{sgn}\left(\mathbf{C}^p - \frac{1}{\mu}\nabla F(\mathbf{C}^p)\right) \tag{8.19}$$

where $\frac{1}{\mu}$ is a step size. We set $\mu^p \in (L, 2L)$, where L is the Lipschitz constant.

Update G.

$$\min F(\mathbf{G}) = ||(\mathbf{B} - \mathbf{C}\,\mathbf{G})\mathbf{W}||_F^2 \tag{8.20}$$

Every column in $\mathbf{G} \in \{0, 1\}^{c\times n}$ represents the hard cluster assignment for sample s (i.e., the vector $\mathbf{g}_s$). It is given by:

$$g_{is}^{p+1} = \begin{cases} 1 & i = \arg\min_k H(\mathbf{b}_s, \mathbf{c}_k^{p+1}) \\ 0 & otherwise \end{cases} \tag{8.21}$$

where $H(\mathbf{b}_s, \mathbf{c}_k)$ is the Hamming distance between the s-th binary code $\mathbf{b}_s$ and the k-th cluster centroid $\mathbf{c}_k$.

- **Step 4: Update the sample weighting matrix W.**
  W is the diagonal sample weight matrix. It is initialized by $w_1 = \ldots = w_s = \ldots = w_n = \frac{1}{n}$. It is updated using the following:

$$\min F(\mathbf{W}) = \sum_{v=1}^{V} \alpha^v(||(\mathbf{B} - \mathbf{U}^v\boldsymbol{\Phi}(\mathbf{X}^v))\mathbf{W}||_F^2) + \lambda||(\mathbf{B} - \mathbf{C}\mathbf{G})\mathbf{W}||_F^2 \qquad s.t. \sum_{s=1}^{n} w_s = 1, w_s > 0, \tag{8.22}$$

The loss function (8.22) is simplified by adopting the following intermediate matrices:
$\mathbf{P}^v = [\mathbf{p}_1^v, \ldots, \mathbf{p}_n^v] = \mathbf{B} - \mathbf{U}^v \boldsymbol{\Phi}(\mathbf{X}^v)$
$\mathbf{M} = [\mathbf{m}_1, \ldots, \mathbf{m}_n] = \mathbf{B} - \mathbf{C}\,\mathbf{G}$

$$F(\mathbf{W}) = \sum_{v=1}^{V} \alpha^v \left(\sum_{s=1}^{n} w_s^2 ||\mathbf{p}_s^v||^2 \right) + \lambda \sum_{s=1}^{n} w_s^2 ||\mathbf{m}_s||^2 - \varepsilon \left(\sum_{s=1}^{n} w_s - 1 \right) \tag{8.23}$$

$$\frac{\partial F(\mathbf{W})}{\partial w_s} = 0 \implies \sum_{v=1}^{V} \alpha^v 2 w_s ||\mathbf{p}_s^v||^2 + 2\lambda w_s ||\mathbf{m}_s||^2 - \varepsilon = 0 \tag{8.24}$$

$$\implies 2\, w_s \left[\sum_{v=1}^{V} \alpha^v \; ||\mathbf{p}_s||^2 + \lambda \; ||\mathbf{m}_s^v||^2 \right] = \varepsilon \tag{8.25}$$

$$\implies 2\, w_s\, A_s = \varepsilon \tag{8.26}$$

where $A_s = \sum_{v=1}^{V} \alpha^v \; ||\mathbf{p}_s^v||^2 + \lambda \; ||\mathbf{m}_s||^2$

$$\implies w_s = \frac{\varepsilon}{2\, A_s} \tag{8.27}$$

$$\sum_{s=1}^{n} w_s = 1 \implies \varepsilon = \frac{1}{\sum_{s=1}^{n} \frac{1}{2\, A_s}} \tag{8.28}$$

$$\implies w_s = \frac{\frac{1}{\sum_{s=1}^{n} (\frac{1}{2 \cdot A_s})}}{2\, A_s} \tag{8.29}$$

- **Step 5: Update the view weight** α^v, $v = 1, \ldots, n$.
  These are initialized by $\alpha^v = \frac{1}{v}$, $\forall v = 1, \ldots, V$.

  With fixed $\mathbf{U}^v$, $\mathbf{B}$, and $\mathbf{W}$, α^v can be optimized using Eq. (8.8). Algorithm 1 summarizes the proposed framework.

8.3.6 Binary Clustering Initialization

We leverage the feature-rich representation of the pretrained visual geometry group model VGG16, recognizing the superior performance of various convolutional neural network (CNN) architectures in detecting object features compared to traditional hand-crafted detectors [11]. Introducing our novel deep approach, bidirectional

Algorithm 1: Auto-weighted binary multi-view clustering via deep initialization (AW-BMVC)

Input:
Multi-view data $\mathbf{X}^v \in \mathbb{R}^{d_v \times n}$, and selected anchors $\mathbf{A}^v \in \mathbb{R}^{d_v \times m}$, $v = 1, \ldots, V$, parameters β, γ, λ, # of clusters c, # of iterations r & t, length of binary codes l.

Output: Binary representation **B**, cluster centroid **C**, cluster indicator **G**.

Initialization: Initialize view weights $\alpha^v = \frac{1}{V}$, initialize sample weights $w_s = \frac{1}{n}$, initialize binary representation **B** (see Sect. 8.3.6).
Compute anchor-based representation $\mathbf{\Phi}(\mathbf{X}^v)$, $v = 1, \ldots, V$ using (8.1).
repeat
  Update $\mathbf{U}^v$ using (8.13).
  Update **B** using (8.15).
  repeat
    Update **C** using (8.19).
    Update **G** using (8.21).
  until *convergence or reach "r" iterations*;
  Update **W** using (8.29).
  Update α using (8.8).
until *convergence or reach "t" iterations*;

fast Fourier transform "BD-FFT," we harness Fourier decomposition for generating effective representative codes [9].

The procedure entails forwarding our image dataset through the VGG16 model and extracting features from the second fully connected (FC) layer, housing 4096 neurons, each attuned to specific features [20]. Subsequently, we create a frequency domain representation by bidirectionally sorting frequencies using FFT. Treating each deep feature vector as a one-dimensional signal, we select coefficients corresponding to "l" low frequencies and transform them into binary codes [9, 10], with the threshold set to the mean of the frequency coefficients.

It is crucial to emphasize that the deep features do not serve as an additional view in the proposed criterion (8.11). Their sole purpose is to facilitate a well-initialized matrix **B**.

8.4 Performance Evaluation

8.4.1 Experimental Setup

8.4.1.1 Datasets

We conducted experiments on four widely used public multi-view image datasets that are commonly employed for benchmarking clustering algorithms: **Caltech101-**

Table 8.2 Datasets used in our experiments. "dim" refers to the feature dimension

Dataset	#Samples	#Views	Feature descriptors	#Classes
Caltech101-7/20	1474/2386	6	48-dim Gabor features	7/20
			40-dim Wavelet moments	
			254-dim Centrist features	
			1984-dim HOG	
			512-dim GIST	
			928-dim LBP	
NUSWIDE-Obj	30,000	5	65-dim Color Histogram	31
			226-dim Color moments	
			145-dim Color correlation	
			74-dim Edge distribution	
			129-dim Wavelet texture	
Scene-15	4485	3	20-dim GIST	15
			59-dim PHOG	
			40-dim LBP	

7, **Caltech101-20**[1] [2], **NUSWIDE-Obj**[2] [19], and **Scene-15**[3] [18]. Multi-view features were extracted to describe each image, and Table 8.2 provides a detailed overview of these datasets.

Caltech101 contains 9,144 images grouped into 101 objects. Following the approach in [22], we selected the widely used object recognition dataset with seven categories, resulting in the Caltech101-7 dataset. Additionally, we selected 2386 images associated with 20 classes, forming the Caltech101-20 dataset.

NUSWIDE-Obj comprises 30,000 images distributed among 31 classes. For this dataset, we utilized five popular descriptors.

Scene-15 consists of 4485 images categorized into 15 indoor and outdoor scenes. Features were extracted from each image, forming three views.

8.4.1.2 Evaluation Metrics and Competitors

We validated the proposed approach using three widely accepted external evaluation criteria [12]: accuracy (ACC), normalized mutual information (NMI), and purity. Our validation involved a thorough comparison with eleven state-of-the-art algorithms, precisely detailed in the related work Sect. 8.2: **RMSC** [15], **DiMSC** [25], **AWP** [3], **WMSC** [32], **BMVC** [31], **OMSC** [26], **LMVSC** [30], **NESE** [29], **GMC** [7], **SMVSC** [14], and **Co-FW-MVFCM** [13].

[1] https://data.caltech.edu/records/20086

[2] https://lms.comp.nus.edu.sg/wp-content/uploads/2019/research/nuswide/NUS-WIDE.html

[3] https://figshare.com/articles/dataset/15-Scene_Image_Dataset/7007177

The compared algorithms were executed with the specified optimal parameter settings from each respective work.

8.4.2 Parameter Sensitivity

The proposed model incorporates three hyperparameters—β, γ, and λ—that are crucial for tuning its behavior and ensuring a stable solution. We conducted an analysis of these parameters, fixing λ at 1e-9 and exploring different values for β and γ from the grid {1e-5, 1e-4, 1e-3, 1e-2 1e-1 2, 4, 6, 10}.

The variability in clustering accuracy across the four datasets for different configurations of β and γ is illustrated in Fig. 8.2.

It is worth noting that the sensitivity across the Caltech101-7/20 datasets is also influenced by the number of selected anchors, recommended to be fewer than 1000 anchors. This recommendation is based on numerical perturbation, which will be further addressed in the convergence analysis (Sect. 8.4.6). Excellent clustering performance is achieved with low values of β ($\beta = 1e - 5$) and relatively high values of γ ($\gamma = 10$). The clustering performance remains relatively stable when

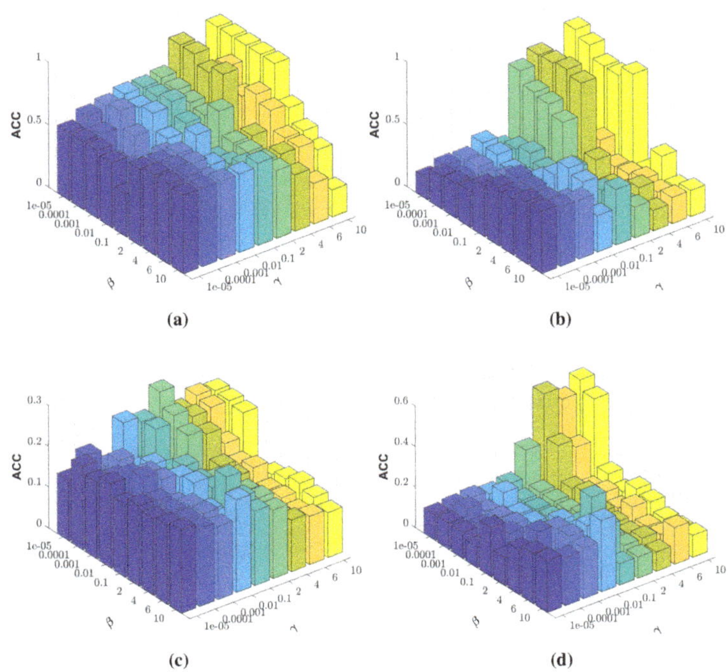

Fig. 8.2 Variability of accuracy with respect to β and γ parameters on: (**a**) Caltech101-7, (**b**) Caltech101-20, (**c**) NUSWIDE-Obj, (**d**) Scene-15

Table 8.3 Best parameter tuning

Datasets	β	γ	λ
Caltech101-7(20)	1e-05	10	1e-09
NUSWIDE-Obj	1e-05	2	1e-09
Scene-15	1e-05	10	1e-09

$1e-5 < \beta < 1e-2$ and $2 < \gamma < 10$. Outside this optimal range, there is a risk of diminishing effectiveness.

The optimal parameter settings for the three parameters are summarized in Table 8.3. Despite the sensitivity mentioned earlier, we achieved excellent clustering results for all tested datasets with only one tuning of γ in a small search range.

8.4.3 Computational Complexity

The proposed AW-BMVC approach effectively addresses the problem of large-scale clustering of multiple views by framing binary code learning as a research challenge. The total complexity of AW-BMVC, considering five optimization operations ($\mathbf{U}^v, \mathbf{B}, \mathbf{C}, \mathbf{G}, \mathbf{W}, \alpha$), is $O(nlm^2Vt)$. This calculation surpasses the complexity of deep feature extraction, which constitutes part of the preprocessing step.

Notably, $l << n$ and $m << n$, and the small number of iterations "t" contributes to the rapid convergence of the proposed model. Consequently, the time complexity can be summarized as $O(n)$, exhibiting linear dependence on n.

The experiments were conducted using Matlab R2019b on a PC machine with a 2.39GHz, i-5-2430M CPU, and 6GB RAM memory. Table 8.4 presents the running time of different methods on the Caltech101-7 dataset.

Among the methods, DiMSC (355.77 sec), OMSC (107.55 sec), LMVSC (135.79 sec), SMVSC (236.32 sec), and Co-FW-MVFCM (1864.57 sec) are relatively time-consuming (more than 100 sec) compared to others. Our method achieves clustering results within 15.23 sec in only $t = 3$ iterations for the Caltech101-7 dataset, with a slight increase in time cost due to the self-weighted term of the sample. AWP, WMSC, and BMVC demonstrate better time efficiency, but AW-BMVC achieves the best clustering results.

Table 8.4 The running time (seconds) of different clustering approaches on the Caltech101-7 dataset

Method	Time (s)	Method	Time (s)
RMSC-2014	92.08	LMVSC-2020	135.79
DiMSC-2015	355.77	NESE-2020	63.18
AWP-2018	7.76	GMC-2020	92.81
WMSC-2018	7.73	SMVSC-2021	236.32
BMVC-2018	6.18	Co-FW-MVFCM-2021	1864.57
OMSC-2019	107.55	AW-BMVC(Ours)	15.23

8.4.4 Ablation Study

Our proposed method comprises three core modules: automatic sample weighting, automatic view weighting, and initialization of binary clustering. In Table 8.5, we present the performance on two tested datasets when each module is either removed or added.

It is important to highlight that the combination of these three indispensable components results in the comprehensive proposal, as indicated by the bold results. The removal of these components leads to the BMVC framework [31] (the first row in Table 8.5).

We observe that the absence of either auto-weighted component or the binary clustering initialization part can lead to a significant drop in performance. Hence, the binary clustering initialization module plays a crucial role in enhancing clustering

Table 8.5 Ablation experimental results. *SAW* sample auto-weighted, *VAW* view auto-weighted, *BCI* binary clustering initialization

	Removing or adding a component			Dataset			
	SAW	VAW	BCI	Caltech101-7	Caltech101-20	NUSWIDE-Obj	Scene-15
ACC	✗	✗	✗	0.2856	0.2355	0.1680	0.2312
NMI				0.1079	0.1864	0.1621	0.1466
Purity				0.5916	0.4392	0.2872	0.2580
ACC	✔	✗	✗	0.2904	0.2921	0.1875	0.1739
NMI				0.1645	0.2149	0.1082	0.1062
Purity				0.6832	0.4715	0.2383	0.1835
ACC	✗	✔	✗	0.3209	0.3814	0.1336	0.2446
NMI				0.2065	0.4932	0.1457	0.2062
Purity				0.7123	0.7318	0.2721	0.2999
ACC	✗	✗	✔	0.5122	0.5159	0.1874	0.5032
NMI				0.4935	0.6830	0.1935	0.43.22
Purity				0.8718	0.8688	0.3081	0.55.74
ACC	✔	✔	✗	0.3141	0.2200	0.1695	0.2881
NMI				0.1583	0.1898	0.1676	0.2080
Purity				0.6784	0.4484	0.2714	0.3097
ACC	✔	✗	✔	0.4274	0.6144	0.1598	0.433
NMI				0.2746	0.4900	0.1552	0.3986
Purity				0.7822	0.6174	0.2811	0.4384
ACC	✗	✔	✔	0.5102	0.4736	0.1853	0.4932
NMI				0.4931	0.6659	0.1957	0.3564
Purity				0.8718	0.8395	0.3137	0.54.62
ACC	✔	✔	✔	**0.9022**	**0.8734**	**0.2190**	**0.5634**
NMI				**0.8733**	**0.8180**	**0.2156**	**0.5089**
Purity				**0.9022**	**0.8873**	**0.322**	**0.5884**

performance. Moreover, attempting to separate either auto-weighted component from the binary clustering initialization variant diminishes the model's capacity.

8.4.5 Clustering Initialization Analysis

To further investigate the impact of different initialization scenarios for the binary codes on our proposed optimization algorithm, we conducted a series of experiments on all datasets using Algorithm 1 with three initialization scenarios for the binary codes matrix $\mathbf{B}$: random binary matrix, nonlinear PCA, and deep FFT.

In the first scenario, a random binary matrix is generated, representing a case where the binary codes are completely independent of the data. The second scenario involves the application of a nonlinear PCA technique adopted by the BMVC method [31]. The third technique, deep FFT, is our proposed initialization scenario.

As shown in Table 8.6, the results obtained with a random binary matrix are noticeably worse, indicating its lack of dependency on the data. The nonlinear PCA method produces homogeneous but suboptimal performance metrics. In contrast, our proposed initialization approach using deep FFT achieves a significant improvement in clustering results, leveraging a new deep feature extraction that explicitly enhances the optimization process.

8.4.6 Convergence Analysis and Effect of the Number of Anchors

Figure 8.3 illustrates the values of the objective function for each iteration on four datasets. The alternating iterative optimization strategy is employed to iteratively update each variable: the mapping matrix $\mathbf{U}^v$, the discrete representation $\mathbf{B}$, the binary cluster centroids $\mathbf{C}$, the cluster indicator matrix $\mathbf{G}$, the auto-weighted sample $\mathbf{W}$, and the auto-weighted view α. The subproblems $\mathbf{U}^v$ and $\mathbf{B}$ originating from Eq. (8.12) and Eq. (8.14) ensure closed-form optimal solutions given by Eq. (8.13) and Eq. (8.15), respectively. The subproblem $\mathbf{C}$ in Eq. (8.17) has an analytical solution using ADPLM [4], as shown in Eq. (8.20). Subsequently, the solution for $\mathbf{G}$ in Eq. (8.21) follows, resembling the K-means learning scheme. The solutions for the automatic weighting of samples in Eq. (8.29) and the automatic weighting of views in Eq. (8.8) represent exact minimum points. Consequently, the loss values of the globally adopted objective function $F(\mathbf{U}; \mathbf{B}; \mathbf{C}; \mathbf{G}; \mathbf{W}; \alpha)$ in Eq. (8.11) decrease rapidly and reach the minimum point after about $t = 5$ iterations, while adhering to the monotonic bound, which serves as a sufficient condition for convergence.

Figure 8.4 illustrates the clustering performance as a function of the number of anchors using the Scene-15 dataset. Clustering performance is visibly affected by this number, with accuracy ranging between 0.4939 ($m = 400$) and 0.5666 ($m = 700$). While there are no specific criteria for determining the optimal number

Table 8.6 Clustering initialization study. *RI* random initialization, *PCA* one-view PCA initialization. *Deep* deep FFT initialization

	Variant											
	RI	PCA	Deep	RI	PCA	Deep	RI	PCA	Deep	RI	PCA	Deep
Dataset	Caltech-7			Caltech-20			NUSWIDE-Obj			Scene-15		
ACC	0.2863	0.2924	**0.9022**	0.2393	0.2200	**0.8734**	0.1508	0.1956	**0.2190**	0.1445	0.2453	**0.5634**
NMI	0.0622	0.2103	**0.8733**	0.1471	0.1898	**0.8180**	0.0785	0.1500	**0.2156**	0.0476	0.1536	**0.5089**
Purity	0.5733	0.6934	**0.9022**	0.4510	0.4484	**0.8873**	0.2041	0.2634	**0.3220**	0.1521	0.2660	**0.5884**

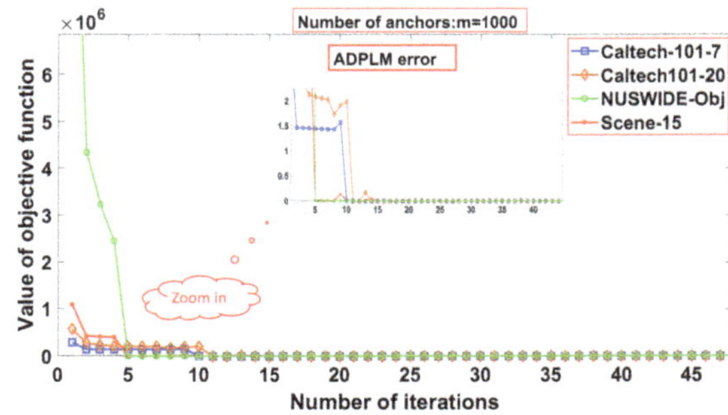

Fig. 8.3 Objective function as a function of iteration number on all datasets. The number of anchors m is set to 1000

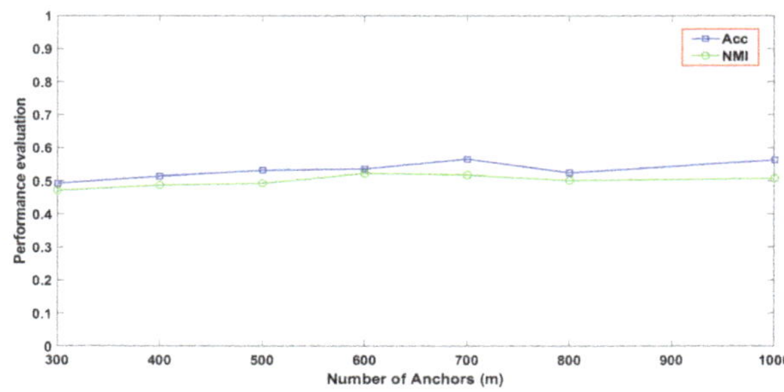

Fig. 8.4 ACC and NMI variation versus the number of anchors on the Scene-15 dataset

of anchors, it is evident that this strongly depends on the number of samples. Taking Scene-15 as an example, if the number of anchors is less than 500, a degradation in clustering performance is observed. However, when m exceeds 500, the performance peaks and stabilizes.

8.4.7 Comparison with State-of-the-Art Multi-view Methods

To assess the effectiveness of the proposed algorithm, extensive experiments were conducted, comparing it with 11 state-of-the-art clustering methods. The performance of each method on four benchmark image datasets (Caltech101-7, Caltech101-20, NUS-WIDE-Obj, and Scene-15) is presented in Table 8.7, with the best-performing method highlighted in bold.

Key observations from the results are as follows:

- OMSC shows unbalanced performance, with extremely low ACC but superior NMI and Purity across all datasets.
- SMVSC and LMVSC achieve the second-best results, albeit with a longer runtime due to their reliance on smooth representation tasks using graph filtering and anchor graph techniques.
- WMSC, AWP, and DiMSC secure the third-best results. WMSC demonstrates lower runtime, emphasizing the minimization of clustering ability between two groups of eigenvectors. DiMSC utilizes Procrustes analysis for improved efficiency, while AWP, being the second most time-consuming, leverages self-expression properties for each view in the original space.

Table 8.7 The clustering performance comparisons on challenging datasets. "–" indicates unavailable results due to out of memory

Methods	Caltech101-7			Caltech101-20			NUSWIDE-0bj			Scene-15		
	ACC	NMI	Purity	ACC	NMI	Purity	ACC	NMI	Purity	ACC	NMI	Purity
RMSC-2014 [15]	0.4037	0.3544	0.8026	0.4035	0.5073	0.7360	0.1473	0.1421	0.2624	0.3482	0.3483	0.3797
DiMSC-2015 [25]	0.5611	0.4221	0.8318	0.4728	0.4935	0.7347	0.1330	0.1363	0.2165	0.2555	0.2083	0.2758
AWP-2018 [3]	0.5685	0.4710	0.8554	0.4953	0.559	0.7594	0.1440	0.1123	0.2446	0.3429	0.3366	0.4035
WMSC-2018 [32]	0.5943	0.496	0.8588	0.5310	0.5893	0.7682	0.1382	0.1344	0.2475	0.4370	0.4341	0.4807
BMVC-2018 [31]	0.2856	0.1079	0.5916	0.2355	0.1864	0.4392	0.1680	0.1621	0.2872	0.2312	0.1466	0.258
OMSC-2019 [26]	0.0257	0.1770	0.9545	0.0255	0.3108	0.9241	0.0678	0.2530	0.4465	0.0084	0.3133	0.8403
LMVSC-2020 [30]	0.7266	0.5193	0.7517	0.5306	0.5271	0.5847	0.1181	0.1063	0.1363	0.3134	0.3297	0.3551
NESE-2020 [29]	0.4857	0.4614	0.8548	0.6085	0.6045	0.7556	–	–	–	0.4312	0.4042	0.4822
GMC-2020 [7]	0.6919	0.6056	0.8846	0.4564	0.3845	0.5549	0.1192	0.1128	0.1205	0.1400	0.1105	0.1464
SMVSC-2021 [14]	0.7354	0.5204	0.8487	0.5692	0.5190	0.6442	0.1254	0.1123	0.1587	0.3583	0.3433	0.3861
Co-FW-MVFCM-2021 [13]	0.4016	0.2819	0.7944	0.3051	0.3887	0.5746	0.1673	0.0913	0.2209	0.2856	0.2822	0.3257
AW-BMVC(Ours)	0.9022	0.8733	0.9022	0.8734	0.8180	0.8873	0.2190	0.2156	0.3220	0.5634	0.5089	0.5884

- NESE proves time-efficient and yields competitive results, benefiting from consistent nonnegative embedding. However, there is room for improvement in addressing the diversity of multiple views explicitly.
- BMVC exhibits computational efficiency but has weaknesses mitigated by our proposed approach, incorporating automatic view weighting, automatic sample weighting, and binary clustering initialization.
- RMSC performs poorly, attributing to its two-step process of consensus graph learning and clustering structure learning.
- Co-FW-MVFCM demonstrates the longest runtime and inadequate feature reduction treatment, resulting in suboptimal performance.

In terms of general clustering metrics, certain baselines outperform others on specific datasets. For instance, BMVC excels for large datasets like NUSWIDE-Obj, WMSC for Scene-15, NESE for Caltech101-20, and GMC for Caltech101-7. Remarkably, AW-BMVC consistently outperforms other methods across all four benchmark image datasets, showcasing its robustness and effectiveness across diverse scenarios.

8.5 Conclusion

In this study, we introduced the auto-weighted binary multi-view clustering (AW-BMVC) via deep initialization method, designed for large-scale multi-view data clustering by optimizing binary clustering through matrix factorization. The inclusion of self-weighted samples and views as a primary component showcases the method's ability to discern between views based on crucial samples, ultimately leading to a comprehensive joint discrete representation. Additionally, a novel deep representation technique was emphasized to address the clustering initialization challenge, yielding positive outcomes for the final clustering performance. The approach demonstrated fast convergence within a few iterations, and empirical results on various datasets established its superiority over several state-of-the-art multi-view clustering methods.

However, for scientific transparency, three inherent limitations should be acknowledged:

1. The reliance on a fixed number of anchors (1000 samples) may lead to selectivity, limiting the model's adaptability to datasets with varying sizes.
2. Despite efforts to enable the model to autonomously learn view and sample weights, the manual tuning of regularization parameters (β, γ, λ) remains necessary for stability, influencing clustering performance.
3. The choice of the pretrained deep Vgg16, as part of the binary matrix initialization technique, confines the evaluation scope to image datasets. Extending the approach to text datasets through diverse binary clustering initialization methods could address this limitation and offer a more comprehensive evaluation.

References

1. Aruna, B.: K-medoids clustering using partitioning around medoids for performing face recognition. Int. J. Soft Comput. Math. Control **3**(3), 1–12 (2014)
2. Fei-Fei, L., Fergus, R. et al.: Learning generative visual models from few training examples: an incremental Bayesian approach tested on 101 object categories. In: 2004 Conference on Computer Vision and Pattern Recognition Workshop, pp. 178–178 (2004). https://doi.org/10.1109/CVPR.2004.383
3. Feiping, N., Lai, T., et al.: Multiview clustering via adaptively weighted procrustes. In: Proceedings of the 24th ACM SIGKDD International Conference on Knowledge Discovery & Data Mining, KDD '18, pp. 2022–2030. Association for Computing Machinery, New York (2018)
4. Fumin, S., Xiang, Z., et al.: A fast optimization method for general binary code learning. IEEE Trans. Image Process. **25**(12), 5610–5621 (2016)
5. Gong, Y., Svetlana, L., et al.: Iterative quantization: a procrustean approach to learning binary codes for large-scale image retrieval. IEEE Trans. Pattern Anal. Mach. Intell. **35**(12), 2916–2929 (2012)
6. Guoqing, C., Shiliang, S., et al.: A survey on multiview clustering. IEEE Trans. Artif. Intell. **2**(2), 146–168 (2021)
7. Hao, W., Yan, Y., et al.: GMC: graph-based multi-view clustering. IEEE Trans. Knowl. Data Eng. **32**(6), 1116–1129 (2019)
8. Hong, T., Chenping, H., et al.: Multiview classification with cohesion and diversity. IEEE Trans. Cybern. **50**(5), pp. 2124–2137 (2018)
9. Jamil, A., Khan, M., et al.: Efficient conversion of deep features to compact binary codes using fourier decomposition for multimedia big data. IEEE Trans. Ind. Inform. **14**(7), 3205–3215 (2018)
10. Jamil, A., Khan, M., et al.: Medical image retrieval with compact binary codes generated in frequency domain using highly reactive convolutional features. J. Med. Syst. **42**(2), pp. 1–19 (2018)
11. Laith, A., Jinglan, Z., et al.: Review of deep learning: Concepts, CNN architectures, challenges, applications, future directions. J. Big Data **8**(1), 1–74 (2021)
12. Liang, N., Z.Yang, et al.: Multi-view clustering by non-negative matrix factorization with co-orthogonal constraints. Knowl.-Based Syst. **194**, 105582 (2020)
13. Miin-Shen, Y., Kristina, S.: Collaborative feature-weighted multi-view fuzzy c-means clustering. Pattern Recogn. **119**, 108064 (2021)
14. Peng, C., Liang, L., et al.: Smoothed multi-view subspace clustering. In: Neural Computing for Advanced Applications, pp. 128–140. Springer, Singapore (2021)
15. Rongkai, X., Yan, P., et al.: Robust multi-view spectral clustering via low-rank and sparse decomposition. In: Proceedings of the AAAI Conference on Artificial Intelligence, vol. 28(1) (2014). https://doi.org/10.1609/aaai.v28i1.8950
16. S. Weglarczyk: Kernel density estimation and its application. ITM Web Conf. **23**, 00037 (2018). https://doi.org/10.1051/itmconf/20182300037
17. Shen, X., Weiwei, L., et al.: Compressed k-means for large-scale clustering. In: Thirty-first AAAI Conference on Artificial Intelligence (2017)
18. S.Lazebnik, Schmid, C., et al.: Beyond bags of features: spatial pyramid matching for recognizing natural scene categories. In: 2006 IEEE Computer Society Conference on Computer Vision and Pattern Recognition (CVPR'06), vol. 2, pp. 2169–2178 (2006). https://doi.org/10.1109/CVPR.2006.68
19. Tat-Seng, C., Jinhui, T., et al.: Nus-wide: A real-world web image database from national university of Singapore. In: Proceedings of the ACM International Conference on Image and Video Retrieval, CIVR '09, pp. 1–9. Association for Computing Machinery, New York (2009)
20. Valdez, A., Megan, P., et al.: Distributed representation of visual objects by single neurons in the human brain. J. Neurosci. **35**(13), 5180–5186 (2015)

21. Van, P., Pham, N., et al.: Multi-view clustering and multi-view models. In: Recent Advancements in Multi-View Data Analytics, pp. 55–96. Springer, Berlin (2022)
22. Wang, B., Xiao, Y., et al.: Robust self-weighted multi-view projection clustering, vol. 34, pp. 6110–6117 (2020)
23. Wang, J., Kumar, S., et al.: Semi-supervised hashing for scalable image retrieval. In: 2010 IEEE Computer Society Conference on Computer Vision and Pattern Recognition, pp. 3424–3431 (2010)
24. Wang, J., Ting, Z., et al.: A survey on learning to hash. IEEE Trans. Pattern Anal. Mach. Intell. **40**(4), 769–790 (2017)
25. Xiaochun, C., Changqing, Z., et al.: Diversity-induced multi-view subspace clustering. In: Proceedings of the IEEE Conference on Computer Vision and Pattern Recognition (CVPR), pp. 586–594 (2015)
26. Xiaofeng, Z., Shichao, Z., et al.: One-step multi-view spectral clustering. IEEE Trans. Knowl. Data Eng. **31**(10), 2022–2034 (2018)
27. Y.Gong, Marcin, P., et al.: Web scale photo hash clustering on a single machine. In: Proceedings of the IEEE Conference on Computer Vision and Pattern Recognition (CVPR), pp. 19–27 (2015)
28. Zhang, Z., Liu, L., et al.: Highly-economized multi-view binary compression for scalable image clustering, pp. 717–732 (2018)
29. Zhanxuan, H., Feiping, N., et al.: Multi-view spectral clustering via integrating nonnegative embedding and spectral embedding. Inform. Fusion **55**, 251–259 (2020)
30. Zhao, K., Wangtao, Z., et al.: Large-scale multi-view subspace clustering in linear time. In: Proceedings of the AAAI Conference on Artificial Intelligence, vol. 34(4), pp. 4412–4419 (2020). https://doi.org/10.1609/aaai.v34i04.5867
31. Zheng, Z., Li, L., et al.: Binary multi-view clustering. IEEE Trans. Pattern Anal. Mach. Intell. **41**(7), 1774–1782 (2018)
32. Zong, L., Zhang, X., et al.: Weighted multi-view spectral clustering based on spectral perturbation. In: Proceedings of the AAAI Conference on Artificial Intelligence, vol. 32(1) (2018). https://doi.org/10.1609/aaai.v32i1.11625

Chapter 9
Clustering with Adaptive Unsupervised Graph Convolution Network

Maria Al Jreidy, Joseph Constantin, Fadi Dornaika, Denis Hamad, and Vinh Truong Hoang

Abstract Graph clustering has become one of the most challenging problems in Deep Learning in recent years. There are a number of methods for classifying nodes, including Graph Convolution Network (GCN), a deep semi-supervised learning method. In this chapter, based on GCN architecture, we propose a deep unsupervised learning scheme. The main contributions are as follows. First, the whole architecture is trained with two unsupervised learning losses based on kernelized features and spectral smoothness. Second, spectral smoothing uses an adaptive and additional graph matrix associated with the predicted soft cluster assignments (node representations) and adaptively integrates additional structure information during the learning phase. The adaptive fused graph used for loss of spectral smoothness takes into account structural information coming from both data features and node deep representations. With the proposed objective function, we are able to develop a powerful graph-based deep clustering. Experiments on four benchmark datasets show that our proposed unsupervised GCN provides better clustering performance than other Graph Neural Network-based techniques for most datasets.

M. Al Jreidy · J. Constantin (✉)
LaRRIS, Faculty of Sciences 2, Lebanese University, Fanar, Lebanon
e-mail: cjoseph@ul.edu.lb

F. Dornaika
University of the Basque Country UPV/EHU, San Sebastian, Spain

IKERBASQUE, Basque Foundation for Science, Bilbao, Spain
e-mail: fadi.dornaika@ehu.eus

D. Hamad
University of the Littoral Opal Coast, Calais, France
e-mail: denis.hamad@univ-littoral.fr

V. T. Hoang
Ho Chi Minh City Open University, Ho Chi Minh City, Viêt Nam
e-mail: vinh.th@ou.edu.vn

F. Dornaika et al. (eds.), *Advances in Data Clustering*,
https://doi.org/10.1007/978-981-97-7679-5_9

Keywords Deep clustering · Graph convolutional networks · Kernel k-means · Spectral clustering · Graph data · Adaptive graph fusion

9.1 Introduction

Graphs, which consist of a set of nodes and the connections between them, are useful for representing challenging and complex problems. The features associated with each node in a graph can be used to describe and represent a variety of real-world applications. To explore the topologies of these networks and find subsets of nodes that group similar nodes together, robust techniques for graph clustering need to be developed. Over the past years, many strategies have been proposed to achieve this goal. During the forward propagation and learning phase, some of them rely only on the graph structure [3, 23], while others rely on the data attributes. For example, the k-means approach [7, 8], which is popular in graph clustering applications due to its simplicity, only considers the data matrix to cluster the graph at the input. The spectral clustering technique, which performs graph clustering based on the graph structure, was then presented [29]. Recently, deep clustering has attracted a lot of interest and has been quite successful in many applications, especially in clustering. The main goal of deep clustering is to eliminate the problem of high-dimensional databases and improve clustering results by combining representation learning and clustering in a single framework. In particular, Convolutional AutoEncoders (CAEs) [17] generally have two layer blocks, one for the encoder and one for the decoder. By minimizing the reconstruction error between its input and its output, which is seen as an image of its input, CAE attempts to identify a representation for each input sample. Deep Clustering Network (DCN) [33] which combines autoencoders with k-means, optimizes the reconstruction loss and the loss of k-means sequentially. Deep Embedded Clustering (DEC) [32] uses Kullback–Leibler (KL) divergence to optimize a self-supervised clustering objective function. Deep Convolutional Embedded Clustering (DCEC) [10] takes advantage of CAE and uses Kullback–Leibler (KL) divergence during the learning phase to optimize the clustering objective function in addition to CAE reconstruction loss. The fundamental drawback of these algorithms is that they use either the data features or the graph structure as input, not both. Thereby, the results they produce are often inadequate. To address this drawback, deep clustering techniques that handle both graph structure and node attributes were then introduced. For example, to learn node embeddings, graph autoencoders (GAEs) [15] and graph variational autoencoders (VGAEs) [15] stack two graph convolutional layers and use loss in reconstructing the graph structure. Additionally, VGAE uses Kullback–Leibler divergence to match the learned embeddings with a Gaussian prior [34]. Since none of these approaches require prior knowledge of basic truth labels during the learning phase, they can all be classified as unsupervised methods.

In contrast, other graph-based algorithms, called semi-supervised techniques, require only a small portion of their inputs to be labeled. Of note is the Graph

Convolution Network (GCN) [15], which has the important advantage of merging the data attributes with the graph structure to produce node representations for classification [14]. However, as mentioned earlier, a GCN often relies on prior knowledge of certain labels that are not available in many databases. Therefore, many works have adopted GCN and its variants for semi-supervised classification and link prediction. Adapting a GCN in an unsupervised way for clustering tasks has several advantages when applying it to real-time data since a GCN takes the attributes of the data and the structure of the graph at its input during the forward propagation phase. By that all sorts of information will be captured from the real-time data to achieve better clustering results. Also, these types of datasets are unlabeled data, which is why an unsupervised technique should be used. For that, a hybrid GCN-based graph clustering model was presented in [12]. Prior to the GCN model, a label sampling mechanism was introduced and used in [30]. These strategies are considered complex because they require a prior label generation procedure to apply the GCN in an unsupervised manner.

The proposed methods mostly adopt graph neural networks to learn node embedding and then apply traditional clustering methods to obtain clusters. However, they usually suffer from the following issues:

1. Some of them rely solely on node features or a graph structure which is unfavorable for clustering due to noise and sparsity problems.
2. Most of the existing clustering methods utilize non-clustering driven losses that cannot well capture the global cluster structure; thus the learned embedding is not sufficient for the downstream clustering task. Moreover, other studies apply semi-supervised learning losses which need labeled data to obtain clusters.

In this chapter, we suggest an unsupervised clustering framework using a GCN architecture allowing deep node representation combining the data attributes with the graph structure. This enables its application with unlabeled data. We present an adaptive framework for learning such an architecture with the aim to cluster the nodes. The main contributions of the chapter are as follows:

- We train a GCN architecture that provides the deep representation of the nodes with two complementary clustering losses. The first loss uses the kernelized features via a kernel k-means clustering loss, where the unknown soft assignments are set to the output of the GCN. The second loss uses information about the adaptive connectivity of the node representations.
- Unlike most GCNs on graphs that assume the graph is given, our proposed method seeks for new graph structures derived from the node representations and the data graph. The proposed method generates an adaptive similarity matrix based on the online estimated node representations from the output of the model during the optimization phase.
- Graph-based regularization loss exploiting node connectivity uses a fused graph obtained by adaptively fusing two graphs: (i) the graph associated with the original data features (fixed graph) and (ii) the graph generated online (associated

with the deep representations of the nodes). Automatic weights are also used to merge these two graphs.

- Experimental results obtained on four public datasets show that the proposed method can significantly improve clustering performance.

The rest of this chapter is organized as follows. Section 9.2 reviews some related work. In Sect. 9.3, we explain the proposed strategy in detail. To demonstrate the effectiveness of the proposed strategy in Sect. 9.4, we give experimental results on a variety of real databases. We also evaluate our clustering approach against other state-of-the-art approaches. Finally, in Sect. 9.5, we draw some conclusions.

9.2 Related Work

The literature on graph clustering can be divided into two categories: structural graph clustering and attributed graph clustering. In structural graph clustering techniques, the nodes are grouped together based on the topology of the graph, such as in deep walk which relies entirely on the graph structure [23]. Based on computing Markov random walks, it generates node sequences using truncated random walks and then obtains node embeddings using the skip-gram model [34]. By optimizing the likelihood that successive nodes will appear at a particular length, node2vec [9] preserves high-order proximity between nodes. A Deep Neural Network (DNN) is used by GraphEncoder [27] to learn the graph similarity matrix. A stacked denoising autoencoder is presented in Deep Neural networks for Graph Representations (DNGRs) [2] to factorize the pointwise mutual information matrix for graph embedding. Adversarial Network Embedding [4], a clustering technique that uses adversarial learning to learn node representations, aims to regularize the embedding learning procedure in order to improve the model robustness and generalizability. This graph clustering approach only takes into account the graph topological structure, ignoring pertinent information from the node attributes.

Both the graph topology and the characteristics of the node are highlighted in attributed graph clustering approaches. For example, in multi-view clustering, Robust Multi-view Spectral Clustering (RMSC) [31] applies a common low-rank transition probability matrix to a traditional Markov chain. Using GAE [15] and VGAE [15], the Adversarially Regularized Variational Graph Autoencoder (ARVGE) [22] learns node embeddings. A semi-supervised method for solving graph classification issues was introduced, and it is known as the GCN [15]. The GCN-based algorithms aid in the fusion of network topology and attribute data, but they require a number of node labels to categorize unlabeled nodes, in contrast to the majority of semi-supervised techniques, which focus on preserving network structure. The Adaptive Multi-channel Graph attentional Convolutional Network (AM-GCN) [28] subsequently generates a feature graph using the KNN method after computing the similarity matrix of nodes from the attribute features of nodes using the cosine or kernel approach. The graph convolution neural network receives both the features and the original graph as input. Self-supervised Graph Convolutional Clustering (SGCC), which combines unsupervised, semi-supervised, and self-supervised views, derives a self-supervised approach based on GCNs for clustering problems using a robust similarity formulation provided by hypergraph-based manifold learning [20]. Additionally, few unsupervised techniques using GCN have been recently developed. Jin et al. [12] proposed an unsupervised model for community discovery using GCN embedding. In order to effectively apply structural information to deep clustering, the Structural Deep Clustering Network (SDCN) algorithm, which combines the advantages of both the autoencoder and the GCN with a unique delivery operator and a dual self-supervised module, was introduced in [5]. In [34], a spectral clustering loss using GCN is used by a Spectral Embedding Network for attributed graph clustering (SENet) to train node embeddings while simultaneously improving the graph structure.

There are two main differences between our proposed method and the SENet approach. First, our approach uses two friendly clustering losses, while the SENet is based only on the spectral clustering loss. Second, our approach uses self-learned structural information (obtained during training) and uses it in the spectral loss function. The SENet approach, on the other hand, relies on the original graph associated with the data.

9.3 Proposed GCN-Based Clustering

Given a graph $G = (V, E, \mathbf{X})$, where $V = \{v_1, v_2, \ldots, v_N\}$ is the set of nodes and $|V| = N$ is the cardinal number of V, E is the set of edges between nodes, and $\mathbf{X} \in R^{N \times d}$ denotes the matrix of node attributes. Moreover, every graph has a well-defined structure described either by the adjacency matrix $\mathbf{A}_{data}$ of $N \times N$ elements with $a_{ij} = 1$ if the two nodes v_i and v_j are connected otherwise $a_{ij} = 0$ or by the similarity matrix $\mathbf{S}$, where s_{ij} is equal to the weight of the edge between nodes v_i and v_j. Note that for graph data (e.g., citation networks), the adjacency matrix is known in advance.

Our goal is to partition the nodes of the graph into C clusters. To achieve this goal, we propose an adaptive model that performs unsupervised clustering of the data by using the GCN blocks that combine the graph structure $\mathbf{A}_{data}$ and the data features at different levels. This architecture, which allows obtaining the node representations, is trained with a hybrid loss that assumes an adaptive graph based on deep node representations during training. The data graph structure is exploited in the GCN blocks, while the adaptive graph is exploited by the regularization loss. The method we propose is called Adaptive Unsupervised Graph Convolution Network (AUGCN). An overview of the proposed method is illustrated in Fig. 9.1.

The remainder of this section is organized as follows. In Sect. 9.3.1, we present some used notations. In Sect. 9.3.2, we describe the GCN architecture used. In Sect. 9.3.3, we explain the generation of an additional graph based on the similarity between node representations. Section 9.3.4 presents the proposed loss function.

Section 9.3.5 describes the spectral loss with a fused graph, and Sect. 9.3.6 describes learning and clustering.

9.3.1 Notations

In this work, a scalar and a vector are identified by lowercase and bold lowercase letters, respectively. A bold uppercase letter is used to indicate a matrix. Table 9.1 provides an overview of the key notations that were used in the chapter.

9.3.2 Model Architecture

GCN becomes the de facto architecture for many graph-based semi-supervised and unsupervised learning tasks. In our work, we adopt a three-layer GCN architecture as shown in Fig. 9.1a. It is composed of two GCN layers, like any other GCN-based method [15], followed by a linear layer. After feeding the matrix of node features and the structure of the graph into the network, represented by $\mathbf{X}$ and $\mathbf{A}_{data}$, respectively, the output of the first layer $\mathbf{H}^1 \in R^{N \times h_1}$ and the second layer $\mathbf{H}^2 \in R^{N \times h_2}$ is calculated as follows:

$$\mathbf{H}^1 = ReLu\,(\widehat{\mathbf{A}}_{data}\,\mathbf{X}\,\mathbf{W}^1) \tag{9.1}$$

$$\mathbf{H}^2 = ReLu\,(\widehat{\mathbf{A}}_{data}\,\mathbf{H}^1\,\mathbf{W}^2) \tag{9.2}$$

Table 9.1 Main notations used in the chapter

Notation	Description
d	Dimension of data
N	Number of data samples
C	Number of clusters
$\mathbf{X} \in R^{N \times d}$	Data matrix
$\mathbf{x}_i \in R^d$	The i-th data sample
$\mathbf{K} \in R^{N \times N}$	Kernel matrix
$\mathbf{A}_{data} \in R^{N \times N}$	Data adjacency matrix
$\mathbf{A}_{out} \in R^{N \times N}$	Similarity matrix of node representations
$\mathbf{I}$	Identity matrix
$\mathbf{D}_{data} \in R^{N \times N}$	Data diagonal matrix
$\mathbf{L}_{data} \in R^{N \times N}$	Data Laplacian matrix
V	Set of nodes
E	Set of edges

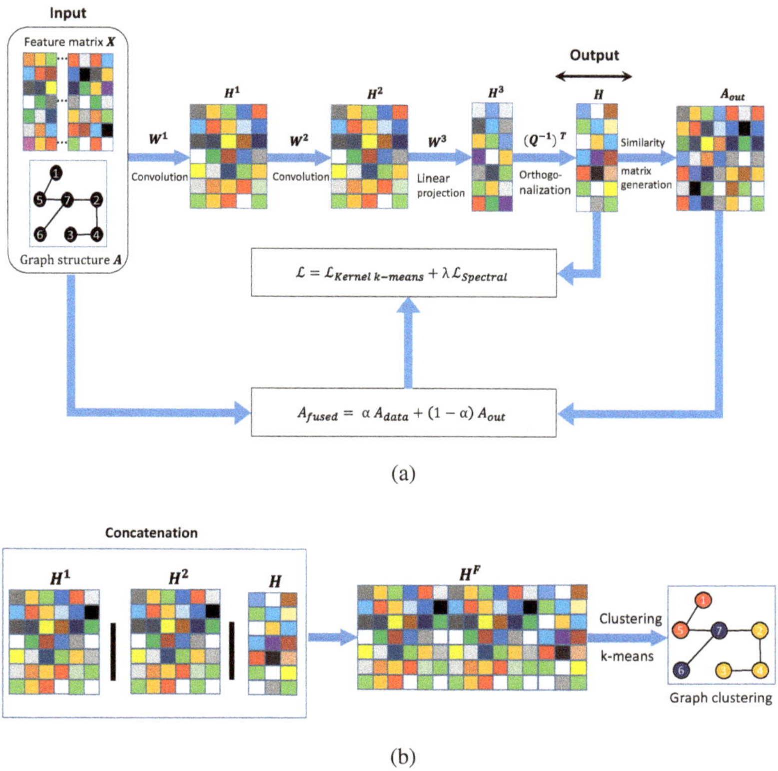

Fig. 9.1 Proposed model: (**a**) architecture and loss function and (**b**) clustering

where $ReLu(.)$ is an activation function given by $ReLu(.) = max(0, .)$ [15]. The trainable matrices of the inner layers are $\mathbf{W}^1 \in R^{d \times h_1}$ and $\mathbf{W}^2 \in R^{h_1 \times h_2}$, while the number of features in the hidden layers is h_1 and h_2, respectively. $\widehat{\mathbf{A}}_{data}$ is the normalized adjacency matrix of the graph calculated as follows:

$$\widehat{\mathbf{A}}_{data} = \widehat{\mathbf{D}}_{data}^{-\frac{1}{2}}(\mathbf{A}_{data} + \mathbf{I})\widehat{\mathbf{D}}_{data}^{-\frac{1}{2}} \tag{9.3}$$

where $\widehat{\mathbf{D}}_{data}$ is the diagonal matrix given by

$$\widehat{D}_{data_{ii}} = \sum_j (\mathbf{A}_{data} + \mathbf{I})_{ij} \tag{9.4}$$

where $\mathbf{I}$ is the identity matrix. In the output layer (the linear layer), we transform $\mathbf{H}^2$ into a C-dimensional space, where C is the number of clusters:

$$\mathbf{H}^3 = \mathbf{H}^2\mathbf{W}^3 \tag{9.5}$$

$\mathbf{W}^3 \in R^{h_2 \times C}$ is also a learnable transformation matrix. Since $\mathbf{H}^3$ plays the role of node representation, it should be orthogonal. For this purpose, we used a Cholesky decomposition [34] on $(\mathbf{H}^3)^T\mathbf{H}^3$:

$$(\mathbf{H}^3)^T\mathbf{H}^3 = \mathbf{Q}^T\mathbf{Q} \tag{9.6}$$

where $\mathbf{Q} \in R^{C \times C}$ is a lower triangular matrix. The orthogonal form of $\mathbf{H}^3$ obtained is given by

$$\mathbf{H} = \mathbf{H}^3\,(\mathbf{Q}^{-1})^T \tag{9.7}$$

We emphasize that the output $\mathbf{H}$ is the node representation provided by the GCN architecture. Although in previous literature, it is considered as the relaxed form of soft cluster assignments.

9.3.3 Similarity Matrix Output

Since the original adjacency matrix may be error-prone, in this work we propose to create a new auxiliary similarity matrix from the output $\mathbf{H}$. This adaptive similarity matrix $\mathbf{A}_{out}$ will encode the pairwise similarity of the node representations (i.e., the rows of $\mathbf{H}$). $\mathbf{A}_{out}$ is computed in the following steps: We compute the N^2 Pearson correlation coefficients between the rows of $\mathbf{H}$ according to Eq. (9.8):

$$\mathbf{A}_{correlation} = \mathbf{H}\,\mathbf{H}^T \tag{9.8}$$

By using the function $ReLu()$, we can make sure that the elements in the resulting matrix $\mathbf{A}_{nonnegative}$ (Eq. 9.9) are greater than or equal to 0.

$$\mathbf{A}_{nonnegative} = ReLu(\mathbf{A}_{correlation}) \tag{9.9}$$

If we use the above dense adjacency matrix, it may rather propagate irrelevant information between nodes. Therefore, we sparsify the affinity matrix by keeping the p highest correlations for each row of $\mathbf{A}_{nonnegative}$, where the p value is a hyperparameter that is usually between 10 and 20. The goal of this step is to obtain the sparse matrix $\mathbf{A}_s$. Finally, we create a symmetric similarity matrix using the following:

$$\mathbf{A}_{out} = \frac{\mathbf{A}_s + \mathbf{A}_s^T}{2} \tag{9.10}$$

The main advantage of $\mathbf{A}_{out}$ is the ability to provide more information about the data structure that is not provided by the original data graph $\mathbf{A}_{data}$. Thus, $\mathbf{A}_{out}$ quantifies the pairwise similarity of the node representations given by the GCN output. Therefore, this additional graph automatically provides complementary information. It is used throughout the training phase of the model by merging it with $\mathbf{A}_{data}$, as shown in Fig. 9.1a and in the next section. At each iteration, the output $\mathbf{H}$ varies and so does the matrix $\mathbf{A}_{out}$.

In the following section, we will elaborate on the loss function of our model, which, as Fig. 9.1a shows, is composed of two loss functions, the kernel k-means loss function, and the spectral loss function.

9.3.4 *Unsupervised Learning Loss*

Our objective is to use GCN in an unsupervised manner to divide the graph nodes into C distinct clusters. In order to accomplish this, we must concentrate on developing an unsupervised loss function that only requires the number of clusters.

The developed loss functions of our proposed AUGCN approach are examined in the following sections. The first loss function exploits data compactness in kernel space, and the second loss function exploits data connectivity based on a new adaptive fused graph.

9.3.4.1 Deep Kernel *k*-Means Loss

The kernel k-means clustering algorithm [6] attempts to recover the soft cluster assignments by minimizing the following objective function:

$$\mathcal{L}_{kernel\ k\text{-}means} = \sum_{j=1}^{C} \sum_{\mathbf{x}_i \in c_j} \|\boldsymbol{\phi}(\mathbf{x}_i) - \mathbf{m}_j\|^2 \tag{9.11}$$

where C is the number of clusters, c_j is the j-th cluster, $\boldsymbol{\phi}(\mathbf{x}_i)$ is a nonlinear function of the original features $\mathbf{x}_i$, and $\mathbf{m}_j$ is the i-th cluster center given by

$$\mathbf{m}_j = \frac{1}{n_j} \sum_{\mathbf{x}_i \in c_j} \boldsymbol{\phi}(\mathbf{x}_i) \tag{9.12}$$

where n_j denotes the number of samples in the cluster c_j.

After some algebraic manipulations, the objective function presented in (9.11) can be expressed as follows:

$$\mathcal{L}_{kernel\ k\text{-}means} = trace(\boldsymbol{\Phi}^T \boldsymbol{\Phi}) - trace(\mathbf{H}^T \boldsymbol{\Phi}^T \boldsymbol{\Phi}\, \mathbf{H}) \tag{9.13}$$

where $trace()$ denotes the trace of a matrix, $\boldsymbol{\Phi} = [\boldsymbol{\phi}(\mathbf{x}_1), \boldsymbol{\phi}(\mathbf{x}_2), \ldots, \boldsymbol{\phi}(\mathbf{x}_i), \ldots, \boldsymbol{\phi}(\mathbf{x}_N)]$, and $\mathbf{H}$ is the $N \times C$ orthonormal cluster assignment matrix, i.e., $\mathbf{H}^T\mathbf{H} = \mathbf{I}$. In the definition of the expression of centers, $\mathbf{H}$ is the hard cluster assignment matrix.

However, when considered as an unknown matrix, it is a soft cluster assignment matrix.

$\boldsymbol{\Phi}$ is the matrix of transformed data represented in kernel space[1] associated with the data. For a clustering problem, the matrix $\mathbf{H}$ is an unknown orthonormal cluster assignment matrix that should be estimated. In our work, this matrix is provided by the GCN architecture. Consider that $\boldsymbol{\Phi}^T\boldsymbol{\Phi}$ is simply the square kernel matrix $\mathbf{K}$ given by

$$K_{ij} = exp(-\frac{1}{\sigma^2}\|\mathbf{x}_i - \mathbf{x}_j\|^2) \tag{9.14}$$

where $\mathbf{x}_i, \mathbf{x}_j$ are data samples, $i, j = 1, \ldots, N$, and σ^2 is set to the average value of the squared distances between the N data points.[2] The loss depicted in Eq. (9.13) becomes as follows:

$$\mathcal{L}_{k\text{-}means} = trace(\mathbf{K}) - trace(\mathbf{H}^T\mathbf{K}\,\mathbf{H}) \tag{9.15}$$

$$= trace\left(\mathbf{K}\,(\mathbf{I} - \mathbf{H}\,\mathbf{H}^T)\right) \tag{9.16}$$

Since $\mathbf{H}$ corresponds to the output of the GCN architecture, the above loss can be called deep kernel k-means loss. Using the k-means objective function helps us exploit the nonlinear attributes induced by the $\mathbf{K}$ matrix to detect the linearly separable clusters in the induced space.

9.3.4.2 Spectral Clustering Loss

The spectral clustering loss enforces the smoothness of the node representations (the rows of the matrix $\mathbf{H}$) over the graph. This is given by the following loss:

$$\mathcal{L}_{spectral} = trace(\mathbf{H}^T\mathbf{L}_{data}\mathbf{H}) \tag{9.17}$$

where $\mathbf{L}_{data}$ is the Laplacian of the original graph matrix $\mathbf{A}_{data}$. In the following, we will propose a new generated adaptive graph that will be used in Eq. (9.17) instead of the original data graph.

9.3.4.3 Global Loss

The total loss that we propose will be simply the combination of the kernel k-means loss and the spectral loss. In this case, our proposed model is trained by minimizing

---

[1] Its explicit form is not needed since only the product $\boldsymbol{\Phi}^T\boldsymbol{\Phi}$ is needed.

[2] In this work, we used the Gaussian kernel. Other types of kernels can also be used.

the following global loss function:

$$\mathcal{L} = trace\left(\mathbf{K}\,(\mathbf{I} - \mathbf{H}\,\mathbf{H}^T)\right) + \lambda\; trace(\mathbf{H}^T\,\mathbf{L}_{data}\mathbf{H}) \tag{9.18}$$

where $\mathbf{H}$ is the orthogonalized output of the model (9.7), and λ is a hyperparameter used to balance between the two terms, usually $\lambda \in [0, 1]$.

We emphasize that the above loss attempts to provide a clustering solution that takes advantage of the compactness of the kernelized samples and the connectivity of the graph used for the spectral clustering loss. This advantage will be quantified in the section of experimental results.

9.3.5 Spectral Clustering Loss with an Adaptive Fused Graph

To capture more structure information about the data during the learning phase, we also consider the fact that the original data graph and the similarity matrix of the output representations (given in Eq. (9.10)) are used to compute the spectral loss, $\mathcal{L}_{spectral}$ (Eq. 9.17). This fused graph is denoted by $\mathbf{A}_{fused}$. Its Laplacian matrix is $\mathbf{L}_{fused}$. This is done by merging $\mathbf{A}_{out}$ and $\mathbf{A}_{data}$ after their normalization by multiplying $\mathbf{A}_{data}$ by $\mathbf{D}_{data}^{-1}$ and $\mathbf{A}_{out}$ by $\mathbf{D}_{out}^{-1}$, where $\mathbf{D}_{data}$ is the degree matrix of $\mathbf{A}_{data}$ and $\mathbf{D}_{out}$ is the degree matrix of $\mathbf{A}_{out}$. Equation (9.19) is used to get the fused graph matrix as convex merging of $\mathbf{A}_{data}$ and $\mathbf{A}_{out}$.

$$\mathbf{A}_{fused} = \alpha\,\mathbf{A}_{data} + (1 - \alpha)\,\mathbf{A}_{out} \tag{9.19}$$

where α is a hyperparameter, $\alpha \in [0, 1]$. In the next section, we present a scheme allowing the automatic computation of α. $\mathbf{L}_{fused}$ will be given by

$$\mathbf{L}_{fused} = \mathbf{D}_{fused} - \mathbf{A}_{fused} \tag{9.20}$$

where $\mathbf{D}_{fused}$ is the degree matrix of $\mathbf{A}_{fused}$. In this case, the proposed loss function given in Eq. (9.18) used to train the model becomes as follows:

$$\mathcal{L} = trace\left(\mathbf{K}\,(\mathbf{I} - \mathbf{H}\,\mathbf{H}^T)\right) + \lambda\; trace(\mathbf{H}^T\,\mathbf{L}_{fused}\mathbf{H}) \tag{9.21}$$

9.3.5.1 Adaptive Fused Graph: Automatic and Adaptive α

In Huang et al. [11], it is stated that learning algorithms are more robust, particularly for unsupervised learning models, when there are fewer parameters to be modified in a model. However, in our proposed model, there is a hyperparameter called α that is used to calculate the value of $\mathbf{A}_{fused}$. Its range of values is 0 to 1, and all of these values must be taken into account in order to achieve the best results.

To avoid this trying and testing procedure, gain more time, and rely less on parameter searching while maintaining strong model performance, we proposed to replace α and $(1-\alpha)$ with automatic weights altering at each epoch. In this instance, we substitute w_{data} and w_{out} for α and $(1-\alpha)$ in the computation of $\mathbf{A}_{fused}$ in Eq. (9.19), respectively. Inspired by the automatic weighting tricks introduced in recent literature (e.g., [11]), these weights are calculated automatically using the following equations:

$$w_{data} = \frac{1}{2\sqrt{(trace(\mathbf{H}^T \mathbf{L}_{data} \mathbf{H}))}} \tag{9.22}$$

$$w_{out} = \frac{1}{2\sqrt{(trace(\mathbf{H}^T \mathbf{L}_{out} \mathbf{H}))}} \tag{9.23}$$

where $\mathbf{H}$ represents the model output, $\mathbf{L}_{data}$ the Laplacian matrix of $\mathbf{A}_{data}$, and $\mathbf{L}_{out}$ the Laplacian matrix of $\mathbf{A}_{out}$.

Then we use Eqs. (9.24) and (9.25) to obtain that their sum is equal to 1.

$$w_{data} = \frac{w_{data}}{w_{data} + w_{out}} \tag{9.24}$$

$$w_{out} = \frac{w_{out}}{w_{data} + w_{out}} \tag{9.25}$$

When $\mathbf{H}$ varies, both weights w_{data} and w_{out} also vary.

9.3.6 *Final Learning and Clustering*

By minimizing the loss function given in Eq. (9.21) with gradient descent, the model parameters $\mathbf{W}^1$, $\mathbf{W}^2$, and $\mathbf{W}^3$ of the AUGCN can be trained. Following model training, we may obtain the hidden node representations $\mathbf{H}^1$, $\mathbf{H}^2$, $\mathbf{H}^3$ and then $\mathbf{H}$ the orthogonalized output.

Through the loss Eq. (9.21), the hidden node representations $\mathbf{H}^1$ and $\mathbf{H}^2$ have implicitly captured the knowledge of the overall cluster structure and the orthogonalized output $\mathbf{H}$. Furthermore, each representation includes information about a particular neighborhood set.

We propose to combine $\mathbf{H}^1$, $\mathbf{H}^2$, and $\mathbf{H}$ to create the final concatenated representation $\mathbf{H}^F \in R^{N \times (h_1+h_2+C)}$, where $\mathbf{H}^F = [\mathbf{H}^1, \mathbf{H}^2, \mathbf{H}]$ in order to take advantage of both the various neighborhood information and knowledge about the global cluster structure. By doing so, the final node representations used for clustering were made rich by exploiting the multistage features obtained by the GCN architecture. The partitions of nodes are then obtained by running k-means on $\mathbf{H}^F$ (see Fig. 9.1b). Algorithm 1 displays the main steps of the AUGCN approach.

Algorithm 1 AUGCN

Input: Attributed Graph $G = (V, \mathbf{E}, \mathbf{X})$, the data adjacency matrix $\mathbf{A}_{data}$, the numbers of highest correlation p, the number of clusters C, the dimensions of inner layers h_1 and h_2, and the hyperparameters λ (α is required for the variant without automatic weighting).
Output: Data partitioning into C clusters.

1. Compute the kernel matrix $\mathbf{K}$ by applying Eq. (9.14).

2. Repeat

- Forward computation of the network layers outputs $\mathbf{H}^1$, $\mathbf{H}^2$, $\mathbf{H}^3$, and $\mathbf{H}$ using Eqs. (9.1), (9.2), (9.5), (9.7).
- Compute the matrix $\mathbf{A}_{out}$ according to Eqs. (9.8) and (9.10).
- Update $\mathbf{A}_{fused}$ using Eqs. (9.24), (9.25) and (9.19).
- Tuning the model parameters $\mathbf{W}^1$, $\mathbf{W}^2$, $\mathbf{W}^3$ by minimizing Eq. (9.21) with gradient descent.

Until the matrix $\mathbf{H}$ does not change or the maximum number of iterations is reached.

3. Concatenate the hidden layers outputs $\mathbf{H}^1$, $\mathbf{H}^2$, and $\mathbf{H}$ learned from the trained AUGCN model to get the complete representation $\mathbf{H}^F = [\mathbf{H}^1, \mathbf{H}^2, \mathbf{H}]$

4. Obtain the clusters by performing k-means on the concatenated output $\mathbf{H}^F$.

9.4 Experiments

In this section, we will provide a quantitative evaluation of the proposed clustering approach using real datasets. We also provide methods comparison and ablation studies.

9.4.1 Datasets

For evaluating the performance of clustering algorithms, all the experiments were run on four datasets: BlogCatalog [16], Cora, Citeseer, and Pubmed [15]. In citation networks like Cora, Citeseer, and Pubmed, nodes stand for publications and are joined by edges when one references another. A feature vector that corresponds to the content of the page binds each node. BlogCatalog (Blog) represents a social network where a node denotes a user, and if two users interact, an edge joins them. Table 9.2 provides a summary of the dataset constitutions.

Table 9.2 Datasets description: Nodes that stand for publications are joined by edges when one publication references another one

Dataset	Nodes	Edges	Features	Clusters
Cora	2708	5429	1433	7
Citeseer	3327	4732	3703	6
Pubmed	19,717	44,338	500	3
BlogCatalog	5196	17,1743	8189	6

9.4.2 Baselines

We compare our method with different baselines:

- Clustering techniques that integrate both node features and graph structure: DeepWalk [23], which uses only graph structure to learn node embeddings, Spectral Clustering [29], which uses the adjacency matrix of the data, and k-means [7], which clusters data based on node features.
- Clustering techniques that integrate both node features and graph structure: GAE and VGAE [15], ARVGE [22], SDCN [5], Self Supervised Clustering based on Manifold Learning and Graph Convolutional Networks (SGCC) [20] and Spectral embedding network (SENET) for attributed graph clustering [34].

9.4.3 Evaluation Metrics and Experimental Setup

The most commonly used evaluation metrics in the deep clustering literature are clustering accuracy (ACC) [1] and normalized mutual information (NMI) [26]. They both are in [0, 1]. The clustering results are better when these values are larger. Equations (9.26) and (9.27) can be used to define ACC and NMI mathematically. In these equations, $\mathbf{y}_{true}$ and $\mathbf{y}_{pred}$ stand for the vectors of ground truth labels and predicted clustering indices, respectively. MI stands for the mutual information function, and En stands for entropy. The Hungarian technique is used to identify the optimal one-to-one mapping, T, connecting the clustering indices to their respective ground truth labels [21].

$$Acc(\mathbf{y}_{true}, \mathbf{y}_{pred}) = \frac{\sum_{i=1}^{N} 1[\mathbf{y}_{true}(i) = T(\mathbf{y}_{pred}(i))]}{N} \tag{9.26}$$

$$NMI(\mathbf{y}_{true}, \mathbf{y}_{pred}) = \frac{MI(\mathbf{y}_{true}, \mathbf{y}_{pred})}{\frac{1}{2}[En(\mathbf{y}_{true}) + En(\mathbf{y}_{pred})]} \tag{9.27}$$

We also use the adjusted Rand index (ARI) [25], along with ACC and NMI, to evaluate our model. The term "RI" in Eq. (9.28), from which ARI is generated, refers to the Rand index, which measures how similar two cluster results are by considering all points found in the same cluster.

$$ARI = \frac{RI - expected(RI)}{max(RI) - expected(RI)} \tag{9.28}$$

The Rand index RI is defined as [24]

$$RI = \frac{a+b}{a+b+c+d} \tag{9.29}$$

where a, b, c, and d are defined for a set $S = \{o_1, o_2, \ldots, o_n\}$ with two partitions $X = \{X_1, X_2, \ldots, X_r\}$ and $Y = \{Y_1, Y_2, \ldots, Y_s\}$ as:

- a, the number of pairs of elements in S that are in the same subset in X and in the same subset in Y
- b, the number of pairs of elements in S that are in different subsets in X and in different subsets in Y
- c, the number of pairs of elements in S that are in the same subset in X and in different subsets in Y
- d, the number of pairs of elements in S that are in different subsets in X and in the same subset in Y

In our experiments, we fit our 2-GCN layers and the third linear layer with randomly initialized weights. We consider $h_1 = h_2 = 16$ as the dimension of the hidden layers [15]. With a learning rate of 0.001, we use the Adam optimizer to train the AUGCN model and run experiments with PyTorch. We set $p = 10$ in the p-highest $\mathbf{A}_{correlation}$ computation step, and λ is set to 0.5. The proposed approach and all competing algorithms were run ten times for 70 epochs on each dataset to ensure the stability of the results [13].

9.4.4 Performance Evaluation

The metrics and their standard deviation results of the proposed and compared algorithms on the datasets are shown in Table 9.3. The best results are displayed in bold. As can be seen, the AUGCN algorithm proposed in this chapter outperforms the baseline algorithms on all datasets except the Cora dataset. For this dataset, AUGCN is the second best method, and SENet is the best algorithm overall. According to the objective of our proposed algorithm, these results can be generally interpreted as follows:

The approach proposed in this chapter, AUGCN, outperforms clustering strategies that use node attributes or graph structure, such as deep walk and spectral clustering, since it effectively integrates and captures both types of information; specifically, in the AUGCN model, the graph structure and attributes are employed for the global loss and feature propagation. For most datasets, AUGCN additionally outperforms approaches that integrate graph structure and data features since it adapts the graph structure and attributes during the feature propagation in the model and the training phase, unlike GAE and VGAE, which only employ them both during the propagation phase. Furthermore, the transmission of information regarding the cluster structure at each layer is facilitated by the use of the kernel

Table 9.3 Clustering results on Cora, Citeseer, Pubmed, and Blog datasets

Method	ACC %	NMI %	ARI %
Cora			
k-means [7]	34.65±1.36	16.73±1.54	9.58±1.07
Spectral [29]	34.19±2.03	19.49±1.94	1.81±1.86
Deep Walk [23]	46.74±2.07	31.75±2.51	23.93±2.04
GAE [15]	53.25±1.85	40.69±1.54	30.47±1.31
VGAE [15]	55.95±0.91	38.45±0.86	29.95±0.83
ARVGE [22]	63.80±0.96	45.00±0.83	37.40±0.71
SDCN [5]	47.03±2.43	25.54±1.92	20.05±1.46
SENet [34]	**71.92±0.66**	**55.08±0.65**	**48.96±1.09**
SGCC [20]	59.22±0.22	39.45±0.34	N/A
AUGCN	71.12±0.35	53.05±0.60	47.50±0.80
Citeseer			
k-means [7]	38.49±2.37	17.02±2.06	12.43±1.92
Spectral [29]	25.91±1.09	11.84±1.68	0.12±1.37
Deep Walk [23]	36.15±1.59	9.66±1.82	10.11±1.75
GAE [15]	41.26±1.79	18.34±1.64	12.93±1.81
VGAE [15]	44.38±1.83	22.71±1.48	15.09±1.56
ARVGE [22]	54.40±1.39	26.10±1.72	24.50±1.64
SDCN [5]	65.96±0.31	38.71±0.32	40.17±0.43
SENet [34]	67.52±0.75	41.70±0.82	42.37±0.97
SGCC [20]	61.11±0.20	32.84±0.24	N/A
AUGCN	**68.76±0.31**	**42.12±0.68**	**43.40±0.65**
Pubmed			
k-means [7]	57.32±3.81	29.12±3.52	25.05±3.46
Spectral [29]	39.74±2.67	3.46±3.09	0.18±2.91
Deep Walk [23]	61.86±2.06	16.71±2.37	18.09±2.54
GAE [15]	64.08±1.66	22.97±1.97	22.25±1.49
VGAE [15]	65.48±1.73	25.09±2.07	24.56±1.86
ARVGE [22]	58.22±1.00	20.62±0.57	20.45±0.65
SDCN [5]	64.20±1.30	22.87±2.04	22.30±2.07
SENet [34]	67.59±0.62	30.61±1.45	29.66±1.13
SGCC [20]	62.95±0.11	26.31±0.07	N/A
AUGCN	**68.47±0.50**	**31.20±1.00**	**31.56±0.89**
Blog			
k-means [7]	21.96±2.94	6.62±3.45	0.15±1.86
Spectral [29]	18.71±2.86	5.78±2.69	0.03±2.51
Deep Walk [23]	36.12±3.65	19.35±3.47	11.32±3.80
GAE [15]	22.09±4.28	9.89±3.71	3.09±3.89
VGAE [15]	23.42±3.47	7.26±2.85	4.32±2.73
ARVGE [22]	39.31±3.05	22.46±2.92	14.55±2.59
SDCN [5]	40.12±2.02	23.32±2.10	15.30±1.47
SENet [34]	41.33±2.16	24.19±1.92	16.14±1.96
SGCC [20]	39.76±2.45	23.01±2.23	N/A
AUGCN	**43.12±0.57**	**24.29±1.81**	**17.50±0.90**

k-means objective loss function and the traditional spectral clustering objective loss function.

Moreover, generating the similarity matrix from the model output at each iteration and using it in the loss function, i.e., in the learning and optimization phase, help us to achieve these results as it can provide more knowledge about the node representation generated from these models. As a result, the generated representation at all layers converges more and more with each epoch to a good representation for the clustering tasks.

9.4.5 Ablation Study

To test how well the proposed approach clusters the data, we perform an ablation study. We run six different scenarios ten times on the Blog dataset. The first scenario is called "AUGCN-spectral" because it relies only on the second term in the loss function shown in Eq. (9.21) after removing the first term $trace\left(\mathbf{K}\left(\mathbf{I}-\mathbf{H}\,\mathbf{H}^{T}\right)\right)$, which represents the kernel k-means loss. In the second scenario, we attempt to obtain $\mathbf{H}$ from the eigenvalue decomposition of C smallest eigenvectors of the matrix $(-\mathbf{K}+\lambda\,\mathbf{L}_{fused})$ instead of the deep GCN, keeping the same loss function as in Eq. (9.21). Since our model in this scenario does not rely on the output of the deep GCN, it is referred to as "A-NoneGCN." To obtain the final clustering results, the third scenario, named "AUGCN-**H**," uses the k-means algorithm only on the model output $\mathbf{H}$. Regarding the fourth scenario, we assume that $\alpha = 1$ in Eq. (9.19); in this case, the loss function will not take into account the information taken from the similarity matrix $\mathbf{A}_{out}$ obtained using the model output. As in this scenario only $\mathbf{A}_{data}$ is used to compute the Laplacian matrix in the second parameter of Eq. (9.21), we will call it AUGCN-$\mathbf{A}_{data}$. In the fifth scenario, named AUGCN-$\mathbf{A}_{out}$, we will only use $\mathbf{A}_{out}$, the adjacency matrix calculated from the model output, to obtain the Laplacian matrix in Eq. (9.21). Hence, during the learning and optimization phase, the adjacency matrix of the initial graph $\mathbf{A}_{data}$ would not be employed. By that, α will be equal to 0 in Eq. (9.19). The sixth scenario is our proposed approach, "AUGCN." The results presented in Table 9.4 led us to the following findings:

1. In order to achieve better results when using a graph clustering algorithm, it is important to use and extract all available information from the graph structure and data features. Due to the fact that "AUGCN-spectral" does not consider the data features encoded in the kernel matrix **K** during the learning phase, our proposed strategy "AUGCN" performs better than this approach.
2. Unlike the "AUGCN-**H**" technique, which only uses **H** during the clustering phase, "AUGCN" uses all the essential information from different layers by concatenating the outputs of $\mathbf{H}^1$, $\mathbf{H}^2$, and **H** layers. As a result, it provides better clustering.

Table 9.4 Ablation study on the Blog dataset

Variant	Omitted component	Omitted variable name	Acc %
AUGCN-spectral	The kernel k-means loss	The kernel matrix **K**	41.10
A-NoneGCN	The GCN architecture	$\mathbf{W}^1, \mathbf{W}^2, \mathbf{W}^3$	39.05
AUGCN-**H**	The first and second layers output	$\mathbf{H}^1$ and $\mathbf{H}^2$	40.10
AUGCN-$\mathbf{A}_{data}$	The adjacency matrix generated from the output	$\mathbf{A}_{out}$	41.70
AUGCN-$\mathbf{A}_{out}$	The adjacency matrix of the data	$\mathbf{A}_{data}$	42.20
AUCGN	None	None	**43.12**

3. The fact that "AUGCN" performs better than "A-NoneGCN" shows that the GCN outputs provide richer information than the eigenvalue decomposition performed in a shallow model.
4. In contrast to "AUGCN-$\mathbf{A}_{data}$" and "AUGCN-$\mathbf{A}_{out}$" approaches, which employ only one adjacency matrix during the learning and optimization process, "AUGCN" captures the important information while using both $\mathbf{A}_{data}$ and $\mathbf{A}_{out}$ matrices in the calculation of the Laplacian matrix used in Eq. (9.21), where $\mathbf{A}_{data}$ and $\mathbf{A}_{out}$ represent the adjacency matrix of the input data and the pairwise similarity of the node representations given by the output, respectively. "AUGCN" produces better results as a consequence. We can also observe that the use of $\mathbf{A}_{out}$ alone provided better performance than using the original data graph $\mathbf{A}_{data}$ alone.

Therefore, all components of the proposed approach contributed to good clustering performance.

9.4.6 *Performance Comparison for Fixed Graph Fusion Versus Adaptive and Automatic Graph Fusion*

In this section, we compare the performance of the proposed method using two different scenarios: (i) The fused graph $\mathbf{A}_{fused} = \alpha\,\mathbf{A}_{data} + (1-\alpha)\,\mathbf{A}_{out}$ is obtained over a fixed α and (ii) the fused graph is obtained over the automatic weights w_{data} and w_{out} as shown in Eqs. (9.24) and (9.25), respectively, i.e., $\mathbf{A}_{fused} = w_{data}\,\mathbf{A}_{data} + w_{out}\,\mathbf{A}_{out}$.

In the first scenario, we tested our algorithm against the four datasets with different values of the coefficient α and determined the clustering accuracy for each of them. The results are shown in Fig. 9.2 after we calculated the average accuracy (over the four datasets) for the different α values. From Fig. 9.2, it can be seen that the highest average accuracy is 58.73% when $\alpha = 0.1$. This means that the best clustering results achieved with fixed fusion the fused graph matrix get 90% pf information from $\mathbf{A}_{out}$ and 10% from $\mathbf{A}_{data}$.

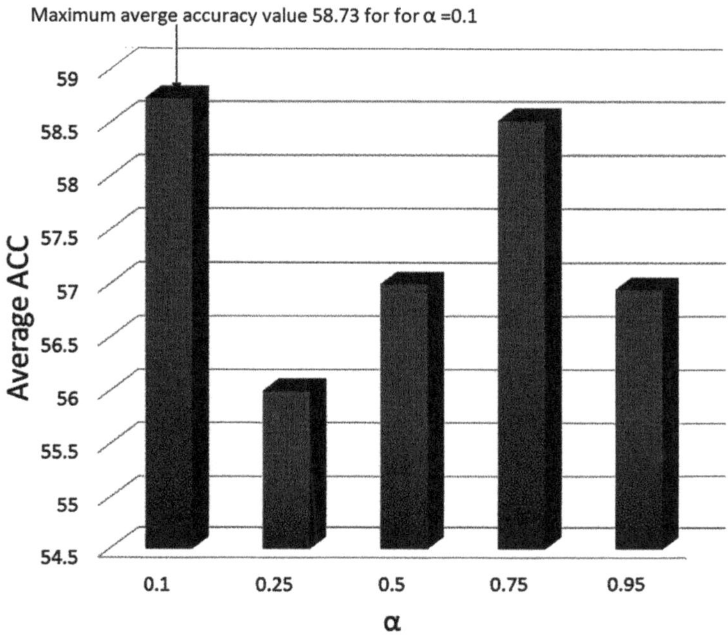

Fig. 9.2 Average accuracy value (four datasets) for different α

We compare the accuracy results obtained with the fixed α with the second scenario where the fused graph is obtained by automatic graph weighting. These are shown in the second column of Table 9.3. The average over the four datasets is 62% when automatic weights are used. This shows that using automatic weights w_{data} and w_{out} instead of fixed α and $(1-\alpha)$ not only eliminates the trial-and-error process of selecting α but also improves the results.

9.4.7 *Impact of* λ *Hyperparameter*

The λ hyperparameter is significant since it is used to balance out the two terms in the suggested loss function (Eq. (9.21)). We examine the impact of λ on the four datasets in this section. Figure 9.3 shows the obtained results of Acc with different values of λ. As we can see, the proposed method gives a satisfactory result for $\lambda = 0.5$ for the four databases. This is why, in our experiments, we have adapted this value of λ in Eq. (9.21).

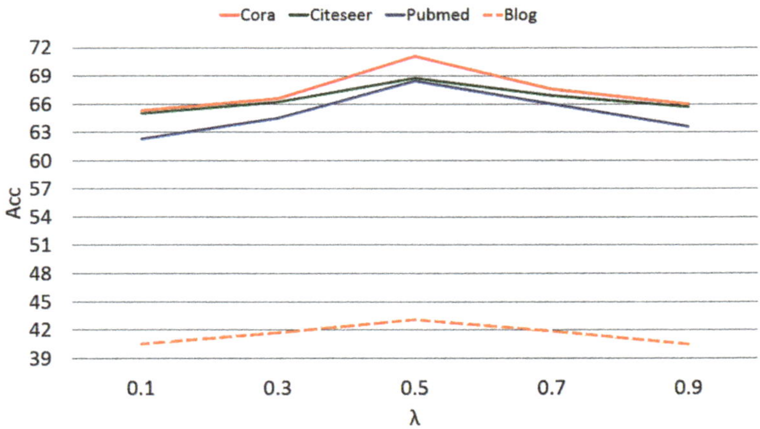

Fig. 9.3 Accuracy (Acc) of the proposed model with different values of λ. The results are obtained on the four databases: Cora, Citeseer, Pubmed, and Blog

9.5 Discussion

Unsupervised machine learning enables machines to reveal patterns that humans might easily miss due to an abundance of data or bias in our thinking process. It explores raw data with an unknown structure, discovering patterns and structures that data scientists would otherwise have no idea about [18]. The advantage of unsupervised machine learning is that it does not require labeled data, which is expensive to manufacture since it requires human experts to identify, categorize, and annotate the data. As such, most of the data are unlabeled. Unsupervised machine learning algorithms, however, can create value from unlabeled data by recognizing previously unknown patterns and discovering features. Graphs are used to represent networked data, which contains complex relationships among individuals and therefore cannot be well represented by traditional vector format. Network applications, like social networks or citation networks, have been developing rapidly in recent years. Consequently, graph learning has also attracted much more attention. Unsupervised graph learning is an important branch of the field since label information is usually not easily accessible [19]. It is much more challenging as unsupervised graph learning aims to model the networked data without training supervision. To perform unsupervised graph learning, previous methods may consider only one aspect of the graph information and use shallow approaches to capture simple or linear relationships among the data. We perform effective graph learning, with graph convolution neural networks in an unsupervised manner. First, we propose a graph convolution neural network, to integrate both node content and graph structure information into a unified framework. We learn node embedding under the guidance of the kernel k-means loss and the spectral clustering loss. We encode the global cluster structure information into the node embedding of different layers. We prove that the final layer can effectively absorb different neighborhood information of each node and well capture the global cluster structure. Experiments on four benchmark datasets show that the performance of the proposed method is better than other compared methods.

9.6 Conclusion

In this work, we focus on the development of an unsupervised adaptive graph convolutional network for clustering. Thanks to the fused adaptive graph matrix created from the original graph and the similarity of the output of the GCN in the learning phase of our proposed approach, we gain a considerable amount of relevant and important knowledge for the node representations, which helps us achieve better clustering results. Moreover, the loss function used in our work combines structural and attribute information about the cluster nodes by adopting two losses that are used separately in reported unsupervised clustering applications. When there is no information about the labels, the proposed loss function can be used to train the GCN model. This method is also a novel adaptive strategy for unsupervised graph clustering that uses a GCN model architecture that takes as input the kernel matrix and the adjacency matrix of the data and adjusts the node features, the graph associated with the spectral loss, during the learning phase. Thanks to the integration of these components, we were able to outperform the state-of-the-art algorithms on the majority of the datasets.

Acknowledgments This work is supported in part by grant PID2021-126701OB-I00 funded by MCIN/AEI/10.13039/501100011033 and by "ERDF A way of making Europe." This work was funded by the University of Littoral Côte d'Opale, France and the Lebanese University.

References

1. Cai, D., He, X., Han, J.: Locally consistent concept factorization for document clustering. IEEE Trans. Knowl. Data Eng. **23**, 902–913 (2011)
2. Cao, S., Lu, W., Xu, Q.: Deep neural networks for learning graph representations. In: The Pacific Rim International Conference on Artificial Intelligence, vol. 30, pp. 1145–1152 (2015). https://doi.org/10.1609/aaai.v30i1.10179
3. Cao, S., Lu, W., Xu, Q.: Deep neural networks for learning graph representations. In: Proceedings of the Thirtieth AAAI Conference on Artificial Intelligence (AAAI), vol. 30, pp. 1145–1152. ACM Press, New York (2016)
4. Dai, Q., Li, Q., Tang, J., Wang, D.: Adversarial network embedding. In: Proceedings of the Thirty-Second AAAI Conference on Artificial Intelligence (AAAI), vol. 32, pp. 2167–2174 (2018). https://ojs.aaai.org/index.php/AAAI/issue/view/301
5. Bo, D., Wang, X., Shi, C., Zhu, M., Lu, E., Cui, P.: Structural deep clustering network. In: Proceedings of the Web Conference, pp. 1400–1410 (2020). https://doi.org/10.1145/3366423.3380214

6. Dhillon, I.S., Guan, Y., Kulis, B.: Kernel k-means: spectral clustering and normalized cuts. In: Proceedings of the Tenth ACM SIGKDD International Conference on Knowledge Discovery and Data Mining, pp. 551–556 (2004). https://doi.org/10.1145/1014052.1014118
7. Fard, M.M., Thonet, T., Gaussier, E.: Deep k-means: jointly clustering with k-means and learning representations. Pattern Recogn. Lett. **138**, 185–192 (2020)
8. Gao, B., Yang, Y., Gouk, H., Hospedales, T.M.: Deep clustering with concrete k-means. In: ICASSP 2020—2020 IEEE International Conference on Acoustics, Speech and Signal Processing (ICASSP), pp. 4252–4256. ACM Press, New York (2020)
9. Grover, A., Leskovec, J.: Node2vec: scalable feature learning for networks. In: Proceedings of the 22nd ACM SIGKDD International Conference on Knowledge Discovery and Data Mining (KDD), pp. 855–864 (2016). https://doi.org/10.1145/2939672.2939754
10. Guo, X., Liu, X., Zhu, E., Yin, J.: Deep clustering with convolutional autoencoders. Lecture Notes in Computer Science (Including Subseries Lecture Notes in Artificial Intelligence and Lecture Notes in Bioinformatics), vol. 10635, pp. 373–382 (2017)
11. Huang, S., Kang, Z., Tsang, I.W., Xu, Z.: Auto-weighted multi-view clustering via kernelized graph learning. Pattern Recogn. **88**, 174–184 (2019). https://doi.org/10.1016/j.patcog.2018.11.007
12. Jin, D., Li, B., Jiao, P., He, D., Shan, H.: Community detection via joint graph convolutional network embedding in attribute network. In: In: Tetko, I., Kůrková, V., Karpov, P., Theis, F. (eds) Artificial Neural Networks and Machine Learning—ICANN 2019: Workshop and Special Sessions. ICANN 2019. Lecture Notes in Computer Science, vol. 11731, pp. 594–606 (2019)
13. Al Jreidy, M., Constantin, J., Dornaika, F., Hamad, D.: Clustering using graph convolution networks. Progr. Artif. Intell. (2024) https://doi.org/10.1007/s13748-023-00310-z
14. Kejani, M.T., Dornaika, F., Talebi, H.: Graph convolution networks with manifold regularization for semi-supervised learning. Neural Netw. **127**, 160–167 (2017)
15. Kipf, T.N., Welling, M.: Semi-supervised classification with graph convolutional networks (2017). arXiv
16. Li, J., Hu, X., Tang, J., Liu, H.: Unsupervised streaming feature selection in social media. In: Proceedings of the 24th ACM International Conference on Information and Knowledge Management (CIKM), pp. 1041–1050 (2015). https://doi.org/10.1145/2806416.2806501
17. Li, F., Qiao, H., Zhang, B., Xi, X.: Discriminatively boosted image clustering with fully convolutional auto-encoders. Pattern Recogn. **83**, 161–173 (2018). https://doi.org/10.1016/j.patcog.2018.05.019
18. Lin, J.S., Guven, E., Duong, L.T., Dinmore, M.D., Hanke, P.A., Magen, B.G., Chavis, J.S.: Unsupervised machine learning by graph analytics on heterogeneous network device data. Proc. Comput. Sci. **140**, 144–151 (2018). https://www.sciencedirect.com/science/article/pii/S1877050918319872
19. Liu, Y., Zheng, Y., Zhang, D., Chen, H., Peng, H., Pan, S.: Towards unsupervised deep graph structure learning (2022)
20. Lopes, L.T., Pedronette, D.C.G.: Self-supervised clustering based on manifold learning and graph convolutional networks. In: Proceedings of the IEEE/CVF Winter Conference on Applications of Computer Vision (WACV), pp. 5634–5643 (2023). https://api.semanticscholar.org/CorpusID:256647249
21. Mrabah, N., Khan, N.M., Ksantini, R., Lachiri, Z.: Deep clustering with a dynamic autoencoder: from reconstruction towards centroids construction. Neural Netw. **130**, 206–228 (2020). https://doi.org/10.1016/j.neunet.2020.07.005
22. Pan, S., Hu, R., Long, G., Jiang, J., Yao, L., Zhang, C.: Adversarially regularized graph autoencoder for graph embedding. In: Proceedings of the Twenty-Seventh International Joint Conference on Artificial Intelligence (IJCAI), pp. 2609–2615 (2018)
23. Perozzi, B., Al-Rfou, R., Skiena, S.: DeepWalk: online learning of social representations. In: Proceedings of the 20th ACM SIGKDD International Conference on Knowledge Discovery and Data Mining. vol. 42, pp. 701–710. ACM Press (2014) https://doi.org/10.1145/2623330.2623732

24. Santos, J.M., Embrechts, M.: On the use of the adjusted rand index as a metric for evaluating supervised classification. Lecture Notes in Computer Science, vol. 5769 (2009)
25. Sinnott, R., Duan, H., Sun, Y.: A case study in big data analytics: exploring twitter sentiment analysis and the weather. In: Big Data: Principles and Paradigms, pp. 357–388 (2016). https://doi.org/10.1016/B978-0-12-805394-2.00015-5
26. Strehl, A., Ghosh, J.: Cluster ensembles: a knowledge reuse framework for combining multiple partitions. J. Mach. Learn. Res. **3**, 583–617 (2002). https://doi.org/10.1162/153244303321897735
27. Tian, F., Gao, B., Cui, Q., Chen, E., Liu, T.Y.: Learning deep representations for graph clustering. In: Proceedings of the Twenty-Eighth AAAI Conference on Artificial Intelligence, vol. 28, pp. 1293–1299 (2014). https://doi.org/10.1609/aaai.v28i1.8916
28. Velickovic, P., Fedus, W., Hamilton, W.L., Liò, P., Bengio, Y., Hjelm, R.D.: Deep graph infomax. In: Proceedings of the 7th International Conference on Learning Representations (ICLR), pp. 1–17 (2019)
29. von Luxburg, U.: A tutorial on spectral clustering. Stat. Comput. **17**, 395–416 (2007). https://doi.org/10.1007/s11222-007-9033-z
30. Wang, X., Li, J., Yang, L., Mi, H.: Unsupervised learning for community detection in attributed networks based on graph convolutional network. Neurocomputing **456**, 147–155 (2021). https://doi.org/10.1016/j.neucom.2021.05.058
31. Xia, R., Pan, Y., Du, L., Yin, J.: Robust multi-view spectral clustering via low-rank and sparse decomposition. In: The Pacific Rim International Conference on Artificial Intelligence, vol. 28, pp. 2149–2155 (2014). https://doi.org/10.1609/aaai.v28i1.8950
32. Xie, J., Girshick, R., Farhadi, A.: Unsupervised deep embedding for clustering analysis. In: Proceedings of The 33rd International Conference on Machine Learning, vol. 48, pp. 478–487 (2016)
33. Yang, B., Fu, X., Sidiropoulos, N.D., Hong, M.: Towards k-means-friendly spaces: Simultaneous deep learning and clustering. In: Proceedings of the 34th International Conference on Machine Learning, vol. 70, pp. 1–10. PMLR (2017)
34. Zhang, X., Liu, H., Wu, X.M., Zhang, X., Liu, X.: Spectral embedding network for attributed graph clustering. Neural Netw. **142**, 388–396 (2021). https://doi.org/10.1016/j.neunet.2021.05.026

Chapter 10
Graph-Based Semi-supervised Learning for Multi-view Data Analysis

Najmeh Ziraki, Fadi Dornaika, Alireza Bosaghzadeh, and Nagore Barrena

Abstract This chapter presents an innovative semi-supervised model that aims to improve data analysis by addressing the limitations of relying on single features. Instead, it advocates the simultaneous use of multiple features to improve performance and avoid misleading conclusions. This algorithm focuses on the joint construction of a consistent graph and label estimation. Key highlights of the proposed method include (1) multi-view data representation: the algorithm processes multi-view data, ensuring a richer and more comprehensive representation; (2) semi-supervised classification: by using both labeled and unlabeled data, the algorithm improves classification accuracy and generalization; and (3) improved visualization: in addition to classification, the method also focuses on improving data visualization to provide better insight and understanding. Experimental results on various image databases show the superiority of the fusion approach over single-feature methods or alternative fusion algorithms. These results underline the importance of leveraging multiple features while creating unified graphs to improve performance in semi-supervised data classification and visualization tasks.

Keywords Multi-view data representation · Semi-supervised classification · Graph fusion · Data visualization

---

N. Ziraki · A. Bosaghzadeh
Faculty of Computer Engineering, Shahid Rajaee Teacher Training University, Tehran, Iran
e-mail: a.bosaghzadeh@sru.ac.ir

F. Dornaika (✉)
University of the Basque Country UPV/EHU, San Sebastian, Spain

IKERBASQUE, Basque Foundation for Science, Bilbao, Spain
e-mail: fadi.dornaika@ehu.eus

N. Barrena
University of the Basque Country UPV/EHU, San Sebastian, Spain
e-mail: nagore.barrena@ehu.eus

F. Dornaika et al. (eds.), *Advances in Data Clustering*,
https://doi.org/10.1007/978-981-97-7679-5_10

10.1 Introduction

Harnessing the power of semi-supervised learning: graph-based techniques for multi-view data fusion and analysis

In today's era of massive data growth, the availability of labeled data is often limited, prompting a surge in interest in semi-supervised learning methodologies. These approaches offer a compelling solution by leveraging both labeled and unlabeled samples to advance machine learning capabilities [10, 20].

Among the array of techniques, graph-based learning has emerged as a prominent methodology capable of exploiting the combined information from labeled and unlabeled data. Central to these algorithms is the construction of affinity graphs, which encode the underlying relationships and proximities between samples, playing a pivotal role in shaping the outcomes of the learning process.

While traditional learning methods predominantly focus on single-feature analysis [7, 8], recent research highlights the advantages of integrating information from multiple features [3, 5, 14, 19, 29]. However, effectively integrating diverse features poses significant challenges, prompting the adoption of graph-based approaches as efficient tools for data fusion and amalgamation.

In the contemporary landscape, there is a growing emphasis on exploring the potential of multi-view learning techniques [11, 16, 28]. These methodologies aim to harness the capabilities of affinity graphs to merge information from different perspectives or "views." Commonly, this involves the integration of multiple preconstructed graphs [2, 12, 27], with variations in fusion methodologies. For example, in [2], researchers employed linear summation for graph fusion, showcasing the diversity of approaches in this domain.

In [6], a novel adaptive weight assignment strategy was introduced, where each view's weight is dynamically determined based on the assumption of label smoothness. This approach allows the algorithm to assign higher weights to views that exhibit more consistent label information across samples, effectively adapting to the varying informativeness of different features. Similarly, in [12], the authors addressed potential noise and misleading information in certain features by incorporating sparsity constraints into the weight assignment process. By imposing these constraints, noisy features are assigned very small weights, effectively excluding them from the fusion process and reducing their influence on the final decision.

Another notable approach, proposed in [27], involves iteratively updating similarity matrices corresponding to each feature type. This method merges these matrices with those from other views while considering similarities among neighboring samples. By iteratively refining the similarity matrices, the algorithm effectively captures the underlying structure of the data across multiple views.

Additionally, some algorithms not only utilize features from the data but also extract features from the sample labels themselves, constructing the affinity matrix based on label similarities [5, 15]. By considering label information in addition to data features, these methods can better capture the intrinsic relationships among samples, leading to improved classification performance.

Recent advancements in multi-view methods have shifted toward directly constructing a unified graph by integrating information from the feature space [17]. By bypassing the need for preconstructed graphs, these approaches can more effectively capture complex relationships among samples across multiple views, leading to enhanced classification accuracy.

In terms of the graph construction process, some algorithms treat it separately from the post-graph learning task, such as label propagation [5, 27]. However, it has been noted that the sample labels can significantly influence the affinity matrix [15]. Therefore, it is preferable to consider graph construction and label propagation as interrelated tasks rather than separate ones, ensuring that the constructed graph accurately reflects the underlying data relationships.

Based on the insights outlined above, this chapter introduces a unified criterion offering several advantages:

- Simultaneous Affinity Graph Construction and Label Propagation: Our approach seamlessly integrates the processes of constructing the affinity graph and propagating labels within a single framework. This streamlines the workflow and ensures consistency between graph construction and label propagation.
- Utilization of Multiple Features: Leveraging multiple features (or descriptors) for each sample enhances the estimation of the unified graph. By incorporating diverse features, our method captures richer relationships between nodes, leading to more robust graph representations.
- Linear Projection Function Estimation: The proposed method includes the estimation of a linear projection function. This function facilitates the prediction of labels for unseen samples, enhancing the algorithm's applicability beyond the training data.
- Elimination of Preconstructed Graphs: Unlike approaches relying on preconstructed graphs from individual views, our method constructs a single graph by integrating features from different views. This simplifies the workflow and avoids the need for separate graph construction steps for each view.

These advantages collectively contribute to a more efficient and effective framework for graph-based semi-supervised learning, enabling improved classification and visualization tasks in multi-view data settings.

Our proposal includes two algorithms: (i) a variant where all views involved in the graph construction are treated equally, and (ii) a variant where these views are represented with learnable weights.

The remainder of this chapter is structured as follows: Sect. 10.2 provides an overview of graph-based semi-supervised learning, outlining fundamental assumptions and discussing previous research in this field. Section 10.3 details our proposed algorithms presenting their methodology and key components. In Sect. 10.4, we present the experimental results obtained from applying our algorithm to public databases, providing insights into its performance and effectiveness. Finally, Sect. 10.5 summarizes the findings and conclusions drawn from this study, discussing potential future directions for research in graph-based semi-supervised learning.

10.2 Related Work

In this section, we introduce the basic notations and definitions used throughout the chapter.

10.2.1 Notations

Let $\mathbf{X} \in \mathbb{R}^{d \times N}$ denote the data matrix, where d represents the dimensionality of each sample and N is the total number of samples. We assume that the first l samples are labeled, and the remaining samples (from $l + 1$ to $l + u$) are unlabeled.

Moreover, $\mathbf{Y} = [\mathbf{Y}_l, \mathbf{Y}_u] = [\mathbf{y}_1; \mathbf{y}_2; \ldots; \mathbf{y}_l; \mathbf{y}_{l+1}; \ldots; \mathbf{y}_{l+u}] \in \mathbb{R}^{N \times C}$ represents the label matrix, where C is the number of classes. For labeled sample $\mathbf{x}_i$ belonging to the cth class, $\mathbf{y}_{ic} = 1$, and it is zero for other elements. For unlabeled samples, the label matrix $\mathbf{Y}_u$ is a zero matrix. Similarly, we define the soft label matrix $\mathbf{F} = [\mathbf{f}_1; \mathbf{f}_2; \ldots.; \mathbf{f}_N] \in \mathbb{R}^{N \times C}$, where F_{ic} represents the probability that sample $\mathbf{x}_i$ belongs to the cth class.

Furthermore, we define an affinity graph as $\mathbf{G} = \{\mathbf{X}, \mathbf{W}\}$, where $\mathbf{X}$ represents the nodes (i.e., the data matrix), and $\mathbf{W}$ is the symmetric affinity matrix. w_{ij} denotes the similarity between nodes $\mathbf{x}_i$ and $\mathbf{x}_j$. Additionally, the Laplacian matrix is defined as $\mathbf{L} = \mathbf{D} - \mathbf{W}$, where $\mathbf{D}$ is a diagonal matrix with $D_{ii} = \sum_{j=1}^{N} W_{ij}$ and zero otherwise.

10.2.2 Gaussian Field and Harmonic Functions [31]

In [31], the authors discuss the cluster assumption, a fundamental concept in label propagation. This assumption posits that nodes close to each other in the feature space should have similar labels. Mathematically, it can be formulated as follows:

$$\min_{\mathbf{F}} \sum_{i,j=1}^{N} ||\mathbf{f}_i - \mathbf{f}_j||^2 \, W_{ij} = \min_{\mathbf{F}} \operatorname{Tr}(\mathbf{F}^T \, \mathbf{L} \, \mathbf{F}) \tag{10.1}$$

where $\operatorname{Tr}(\cdot)$ indicates the trace of a matrix. In this equation, $\mathbf{f}_i$ and $\mathbf{f}_j$ represent the soft labels of sample $\mathbf{x}_i$ and $\mathbf{x}_j$ respectively, and W_{ij} denotes the similarity between them.

Equation (10.1) asserts that if the similarity W_{ij} between samples $\mathbf{x}_i$ and $\mathbf{x}_j$ is high, indicating that they are similar, then their corresponding soft labels $\mathbf{f}_i$ and $\mathbf{f}_j$ should also be similar, as reflected by the small difference between them.

10.2.3 Local and Global Consistency [30]

Another important consideration is label fitness, which ensures that the predicted labels of labeled samples closely match their true labels. Mathematically, this can be expressed as follows:

$$\min_{\mathbf{F}} \sum_{i=1}^{l} ||\mathbf{f}_i - \mathbf{y}_i||^2 = \min_{\mathbf{F}} \mathrm{Tr}\,(\mathbf{F} - \mathbf{Y})^T\, \mathbf{U}\,(\mathbf{F} - \mathbf{Y}). \tag{10.2}$$

Here, the diagonal matrix $\mathbf{U}$ is defined such that $U_{ii} = 0$ for unlabeled samples and $U_{ii} = 1$ for labeled samples.

By incorporating Eq. (10.2) into Eq. (10.1), a new objective function, known as local and global consistency [30], can be defined as:

$$\min_{\mathbf{F}} \sum_{i,j=1}^{N} ||\frac{1}{\sqrt{D_{ii}}}\mathbf{f}_i - \frac{1}{\sqrt{D_{jj}}}\mathbf{f}_j||^2\, W_{ij} + \mu \sum_{i=1}^{l} ||\mathbf{f}_i - \mathbf{y}_i||^2 \tag{10.3}$$

Here, μ serves as the balance parameter.

10.2.4 Review on Flexible Manifold Embedding [21]

A limitation of the approaches described in [31] and [30] is their inability to handle new samples for label prediction. When a new sample arrives, a new graph must be constructed using the new sample, and the minimization function must be solved again.

To address this limitation, the flexible manifold embedding (FME) method [21] introduces a projection matrix calculation to its minimization function. This projection matrix enables the estimation of labels for arbitrary samples through linear mapping, effectively defining a function to map data features to label space.

The objective function of FME is defined as follows:

$$\min_{\mathbf{F},\mathbf{Q},\mathbf{b}} \mathrm{Tr}\,(\mathbf{F}^T\, \mathbf{L}\, \mathbf{F}) + \mathrm{Tr}\,(\mathbf{F} - \mathbf{Y})^T\, \mathbf{U}\,(\mathbf{F} - \mathbf{Y}) + \mu\,(\gamma\, ||\mathbf{X}^T \mathbf{Q} + \mathbf{1}\mathbf{b}^T - \mathbf{F}||^2 + ||\mathbf{Q}||^2) \tag{10.4}$$

Here, the first term penalizes pairs of samples with high similarity but different labels, known as the label smoothness term. The second term represents the fitness of labels on labeled samples. The third term consists of two components: the error of label fitness for all samples adopting linear projection and the regularization of the projection matrix. μ and γ are balance parameters, and $||.||^2$ denotes the ℓ_2 norm of a matrix.

The solution to Eq. (10.4) has a closed form as follows:

$$\mathbf{Q} = \gamma\,[\gamma\,\mathbf{X}\mathbf{H}_c\mathbf{X}^T + \mathbf{I}]^{-1}\mathbf{X}\mathbf{H}_c\mathbf{F} \tag{10.5}$$

$$\mathbf{b} = \frac{1}{l+u}[\mathbf{F}^T\mathbf{1} - \mathbf{W}^T\mathbf{X}\mathbf{1}] \tag{10.6}$$

$$\mathbf{F} = (\mathbf{U} + \mathbf{L} + \mu\,\gamma\,\mathbf{H}_c - \mu\,\gamma^2\,\mathbf{M})^{-1}\mathbf{U}\,\mathbf{Y} \tag{10.7}$$

where $\mathbf{M} = \mathbf{X}^T\,\mathbf{X}\,(\gamma\,\mathbf{X}^T\,\mathbf{X} + \mathbf{I})^{-1}.\ \mathbf{X}$, and $\mathbf{H}_c = \mathbf{I} - \frac{1}{l+u}\mathbf{1}\,\mathbf{1}^T$ are the centered data matrix and the centering matrix, respectively.

The mathematical derivations for Eqs. (10.5), (10.6), and (10.7) are provided in [21].

10.2.5 Data Smoothness Assumption

The data smoothness assumption posits that samples close to each other in the embedding space should exhibit similar labels. Mathematically, this can be expressed using the Laplacian matrix **L** of the affinity matrix **P** as follows:

$$\min_{\mathbf{S}} \frac{1}{2} \sum_{i,j=1}^{N} ||\mathbf{x}_i - \mathbf{x}_j||^2\, P_{ij} = \min_{\mathbf{S}} \mathrm{Tr}(\mathbf{X}\,\mathbf{L}\,\mathbf{X}^T) \tag{10.8}$$

Here, **X** represents the data matrix, and **L** is the Laplacian matrix of the graph **P**.

10.3 Proposed Approach

In this section, we present two semi-supervised models.

The first model assumes that all views contribute equally to the learning process. This approach treats each view with the same level of importance, integrating them into a unified model without differentiating their influence.

The second model introduces a weighted learning approach, where the weights of each view are also learned during the training process. This method allows the model to assign different levels of importance to each view based on their contribution to the overall learning task, potentially enhancing performance by emphasizing more informative views.

10.3.1 Equally Weighted Model

10.3.1.1 Learning Model

We propose a joint graph construction and multi-view flexible manifold embedding (JGC-MFME) algorithm by incorporating the data smoothness assumption into the FME framework. This approach offers two key advantages:

First, it allows for the direct estimation of affinity graphs, eliminating the need for preconstructed graphs. This simplifies the process and ensures that the graph construction aligns with the multi-view learning framework.

Second, it enables the utilization of multiple sources of information to construct a graph that effectively captures the relationship between samples. This enhances the robustness of the graph representation.

Additionally, our proposed method unifies the graph construction and classification into a single criterion, streamlining the optimization process.

To adapt the data smoothness term for multi-view learning, we redefine it as follows:

$$\min_{\mathbf{S}} \sum_{v=1}^{V} \sum_{i,j=1}^{N} ||\mathbf{x}_{vi} - \mathbf{x}_{vj}||^2 \, S_{ij} = \min_{\mathbf{S}} \sum_{v=1}^{V} \mathrm{Tr}(\mathbf{X}_v \, \mathbf{L} \, \mathbf{X}_v^T) \tag{10.9}$$

Here, $\mathbf{x}_{vj}$ represents the vth view of the jth sample, and V denotes the number of views. This formulation ensures that the unified similarity matrix satisfies the data smoothness assumption across all views.

The proposed criterion is constructed by incorporating Eq. (10.9) into the FME framework as follows:

$$\begin{aligned} \min_{\mathbf{S},\mathbf{F},\mathbf{Q},\mathbf{b}} \; & \Big[\sum_{v=1}^{V} \mathrm{Tr}(\mathbf{X}_v \, \mathbf{L} \, \mathbf{X}_v^T) + \lambda \, \mathrm{Tr}(\mathbf{F}^T \, \mathbf{L} \, \mathbf{F}) + \\ & \mathrm{Tr}((\mathbf{F} - \mathbf{Y})^T \, \mathbf{U} \, (\mathbf{F} - \mathbf{Y})) + \gamma \, ||\mathbf{S}||^2 + \\ & \mu \, (||\mathbf{Q}||^2 + \alpha \, ||\mathbf{X}_{conc}^T \mathbf{Q} + \mathbf{1}\mathbf{b}^T - \mathbf{F}||^2) \Big] \\ & \text{s.t.} \quad 0 \le S_{ij} \le 1, \quad \sum_{j=1}^{N} S_{ij} = 1 \end{aligned} \tag{10.10}$$

In Eq. (10.15), $\mathbf{S}$ represents the unified similarity matrix, $\mathbf{F}$ is the soft label matrix, $\mathbf{Q}$ is the projection matrix, and $\mathbf{b}$ is the bias vector. The parameters λ, γ, μ, and α are balance parameters, and $||.||^2$ denotes the ℓ_2 norm of a matrix.

Notably, since we have multiple features, we concatenate them to form the matrix $\mathbf{X}_{conc}$. Consequently, the size of $\mathbf{X}_{conc}$ becomes $(\sum_{v=1}^{V} d^v) \times N$.

10.3.1.2 Optimization

Below, we outline an iterative alternating scheme to solve Eq. (10.15):

Step 1: Fix F, Q, and b and update S
When **F**, **Q**, and **b** are fixed, Eq. (10.15) simplifies to:

$$\begin{aligned} \min_{\mathbf{S}} \quad & \sum_{v=1}^{V} \mathrm{Tr}\,(\mathbf{X}_v\, \mathbf{L}\, \mathbf{X}_v^T) \;+ \lambda\, \mathrm{Tr}(\mathbf{F}^T\, \mathbf{L}\, \mathbf{F}) + \gamma\, ||\mathbf{S}||^2 \\ \text{s.t.} \quad & 0 \leq S_{ij} \leq 1, \quad \sum_{j=1}^{N} S_{ij} = 1 \end{aligned} \tag{10.11}$$

Here, d_{ij}^x represents the distance between nodes $\mathbf{x}_i$ and $\mathbf{x}_j$ in all views, and d_{ij}^f represents the distance between their soft labels.

We rewrite Eq. (10.11) as:

$$\begin{aligned} \min_{\mathbf{S}} \quad & \sum_{i,j=1}^{N} S_{ij}\, (d_{ij}^x + \lambda\, d_{ij}^f) + \gamma\, ||\mathbf{S}||_2^2 \\ = \min_{\mathbf{S}} \quad & \sum_{i,j=1}^{N} [S_{ij}\, d_{ij} + \gamma\, S_{ij}^2] \\ \text{s.t.} \quad & 0 \leq S_{ij} \leq 1, \quad \sum_{j=1}^{N} S_{ij} = 1 \end{aligned} \tag{10.12}$$

The solution to Eq. (10.18) can be obtained using the algorithm presented in [17].

Step 2: Fix S and update Q, b, and F
With **S** fixed, Eq. (10.15) simplifies to:

$$\begin{aligned} \min_{\mathbf{F},\mathbf{Q},\mathbf{b}} \quad & \lambda\, \mathrm{Tr}(\mathbf{F}^T\, \mathbf{L}\, \mathbf{F}) \;+ (\mathbf{F} - \mathbf{Y})\, \mathbf{U}\, (\mathbf{F} - \mathbf{Y})^T + \\ & \mu\, (||\mathbf{Q}||^2 + \alpha\, ||\mathbf{X}_{conc}^T \mathbf{Q} + \mathbf{1}\mathbf{b}^T \;- \mathbf{F}||^2) \end{aligned} \tag{10.13}$$

This is equivalent to the FME problem shown in Eq. (10.4). Therefore, the solution can be obtained using Eqs. (10.5), (10.6), and (10.7).

It is worth noting that the label of a test sample can be estimated using:

$$\mathbf{f} = \mathbf{X}_{conc}^T\, \mathbf{Q} + \mathbf{1}\mathbf{b}^T \tag{10.14}$$

Here, $\mathbf{X}_{conc}$ is the concatenation of features from all views.

10.3.2 View-Weighted Model

10.3.2.1 Learning Model

On the other hand, we now present a variant of the presented method in which the various integrated views are not treated equally, but are given an added weight (JGC-MFME Weighted). Thus, the equation presented in Eq.10.11 is modified in order to take into account the weight corresponding to each view (Eq. 10.15 where β_v^p represents the weight of view v).

$$\min_{\beta,\mathbf{S},\mathbf{F},\mathbf{Q},\mathbf{b}} \Big[\sum_{v=1}^{V} \beta_v^p \, Tr(\mathbf{X}_v \, \mathbf{L} \, \mathbf{X}_v^T) + \lambda \, Tr(\mathbf{F}^T \, \mathbf{L} \, \mathbf{F}) + Tr((\mathbf{F}-\mathbf{Y})^T \, \mathbf{U} \, (\mathbf{F}-\mathbf{Y}))$$
$$+\gamma \, ||\mathbf{S}||^2 + \mu \, (||\mathbf{Q}||^2 + \alpha \, ||\mathbf{X}_{conc}^T \mathbf{Q} + \mathbf{1}\mathbf{b}^T - \mathbf{F}||^2) \Big]$$
$$\text{s.t.} \quad 0 \le S_{ij} \le 1, \quad \sum_{j=1}^{N} S_{ij} = 1 \qquad (10.15)$$

10.3.2.2 Optimization

To solve the problem presented in Eq. 10.15 we propose an iterative alternate solution. We proceed as follows:

Fix S and $\boldsymbol{\beta}$ and calculate Q, b, and F The minimization problem of Eq. (10.15) is reduced to:

$$\min_{\mathbf{F},\mathbf{Q},\mathbf{b}} \; \lambda \, Tr(\mathbf{F}^T \, \mathbf{L} \, \mathbf{F}) + (\mathbf{F}-\mathbf{Y}) \, \mathbf{U} \, (\mathbf{F}-\mathbf{Y})^T + \mu \, (||\mathbf{Q}||^2 + \alpha \, ||\mathbf{X}_{conc}^T \mathbf{Q} + \mathbf{1}\mathbf{b}^T - \mathbf{F}||^2) \qquad (10.16)$$

The above is the FME formulation in 10.4

Consequently, the Eqs. (10.5), (10.6), and (10.7) can be adopted to calculate **Q**, **b**, and **F**, respectively.

Fix F, Q, b, and $\boldsymbol{\beta}$ and calculate S The proposed objective function in Eq. (10.15) can be simplified to

$$\min_{\mathbf{S}} \; \sum_{v=1}^{V} \beta_v^p \, Tr \, (\mathbf{X}_v \, \mathbf{L} \, \mathbf{X}_v^T) + \lambda \, Tr(\mathbf{F}^T \, \mathbf{L} \, \mathbf{F}) + \gamma \, ||\mathbf{S}||^2 \; \text{s.t.}$$
$$0 \le S_{ij} \le 1, \quad \sum_{j=1}^{N} S_{ij} = 1 \qquad (10.17)$$

which can be written as

$$\begin{aligned} \min_{\mathbf{S}} \quad & \sum_{v=1}^{V} \beta_v^p \sum_{i,j=1}^{N} ||\mathbf{x}_i^v - \mathbf{x}_j^v||_2^2 \, S_{ij} + \lambda \sum_{i,j=1}^{N} ||\mathbf{f}_i - \mathbf{f}_j||_2^2 S_{ij} + \gamma \, ||\mathbf{S}||_2^2 \\ \equiv \min_{\mathbf{S}} \quad & \sum_{i,j=1}^{N} S_{ij} \left(\sum_{v=1}^{V} \beta_v^p \, ||\mathbf{x}_i^v - \mathbf{x}_j^v||_2^2 + \lambda \, ||\mathbf{f}_i - \mathbf{f}_j||_2^2 \right) + \gamma \, ||\mathbf{S}||_2^2 \\ \equiv \min_{\mathbf{S}} \quad & \sum_{i,j=1}^{N} [S_{ij} \, d_{ij} + \gamma \, S_{ij}^2] \quad \text{s.t.} \quad 0 \le S_{ij} \le 1, \quad \sum_{j=1}^{N} S_{ij} = 1 \end{aligned} \tag{10.18}$$

where $d_{ij} = \sum_{v=1}^{V} ||\mathbf{x}_i^v - \mathbf{x}_j^v||_2^2 + \lambda \, ||\mathbf{f}_i - \mathbf{f}_j||_2^2$. Since Eq. (10.18) is independent between different i, we can deal with it individually for each row $\mathbf{s}_i$ as (the vector $\mathbf{d}_i$ is set to $(d_{i1}, \ldots, d_{iN})^T$):

$$\min_{\mathbf{s}_i} \sum_{j=1}^{N} S_{ij} \, d_{ij} + \gamma \, S_{ij}^2 \equiv \min_{\mathbf{s}_i} ||\mathbf{s}_i + \frac{1}{2\gamma} \mathbf{d}_i||_2^2 \quad \text{s.t.} \quad 0 \le S_{ij} \le 1, \quad \sum_{j=1}^{N} S_{ij} = 1 \tag{10.19}$$

A closed-form solution for $\mathbf{s}_i$ $(i = 1, \ldots, N)$ can be found in [17].

Fix F, Q, b, and S and calculate $\boldsymbol{\beta}$ Eq. (10.15) can be simplified as

$$\min_{\boldsymbol{\beta}} \sum_{v=1}^{V} \beta_v^p \, C_v + \eta \, (\sum_{v=1}^{V} \beta_v - 1) \tag{10.20}$$

where $C_v = Tr(\mathbf{X}_v^T \mathbf{L} \mathbf{X}_v)$ and η is the Lagrangian multiplier. To estimate the weight coefficients (i.e., β_v, $v = 1, \ldots, V$), we vanish the derivative of Eq. (10.20) w.r.t. β_v and use $(\sum_{v=1}^{V} \beta_v = 1)$. Finally, we get:

$$\beta_v = \frac{\frac{1}{C_v^{\frac{1}{p-1}}}}{\sum_{v=1}^{V} (\frac{1}{C_v})^{\frac{1}{p-1}}} \tag{10.21}$$

Once the model is estimated (i.e., **F**, **Q**, **b**, and **S** are recovered), the label (row vector) of any test sample can be predicted via the following mapping:

$$\mathbf{f} = \mathbf{x}_{conc}^T \mathbf{Q} + \mathbf{b}^T \tag{10.22}$$

where $\mathbf{x}_{conc}$ is the vector constructed by concatenating the features of the views V views.

10.4 Experimental Results

In this section, we compare the performance of our proposed method with several state-of-the-art algorithms, including FME with single feature, FME with concatenated features, similarity network fusion (SNF) [27], sparse multiple graph integration (SMGI) [12], dynamic graph fusion label propagation (DGFLP) [15], multi-view local and global consistency (MLGC) [2], and auto-weighted multiple graph learning (AMGL) [18].

10.4.1 Datasets

We evaluate the performance using four image databases: PF01 [4], multiple pose, illumination, expressions (PIE) [24], The facial recognition technology (FERET) [22], and Modified National Institute of Standards and Technology (MNIST) [13]. The PF01 database contains 1819 images from 103 subjects, with 17 images each. For the PIE database, we use a subset of 1926 images from 68 subjects. In the FERET database, we select a subset consisting of 1400 images of 200 subjects, with each subject having 7 images. The MNIST database consists of 70,000 handwritten digit images, with 60,000 images for training and 10,000 for testing.

10.4.2 Image Descriptors

For multi-view methods, we need to extract multiple features from the images in the databases. For face databases, we extract local binary patterns (LBP) [1], Gabor [23], and covariance (Cov) [26] as image features. After resizing the images to 32×32 pixels, we apply the LBP operator with a radius of one and a neighborhood size of eight, resulting in a 900-feature vector. The second feature is the Gabor wavelet [23], applied at five scales and eight orientations, resulting in a 2560-feature vector. The third feature is the covariance feature [26], where each image is divided into nine blocks, and the results are concatenated into a 405-feature vector.

For the MNIST database, given its size, we use deep networks to extract features as they provide better representations. We input the images to a deep network and extract information from the layer before the last layer. We utilize two networks, VGG-16 and VGG-19 [25]. Each deep feature is a vector with 4096 elements.

Table 10.1 provides a summary of the databases used and the features extracted.

In the preprocessing step, we apply principal component analysis (PCA) to reduce the dimensionality of the features, preserving 90% of the energy. Then, each feature is normalized to have zero mean and unit variance.

Next, we split the datasets into training and test data. In the semi-supervised context, the training set is further divided into labeled and unlabeled sets. For each

Table 10.1 Short description of the adopted databases

Dataset	#Class	#Samples	#Samples in each class	Extracted features	Dimension
PF01	107	1819	17	Cov [26]	405
				Gabor [23]	2560
				LBP [1]	900
PIE	68	1926	27~29	Cov [26]	405
				Gabor [23]	2560
				LBP [1]	900
FERET	200	1400	7	Cov [26]	405
				Gabor [23]	2560
				LBP [1]	900
MNIST	10	60000	5421~6742	VGG16 [25]	4096
				VGG19 [25]	4096

training dataset, we choose l samples as labeled and the rest as unlabeled. We repeat the experiments for ten combinations of labeled and unlabeled samples, resulting in ten splits.

To determine the parameters for the proposed algorithm, we use the first split of labeled/unlabeled data and try various values. The parameter yielding the highest accuracy is chosen as the optimal parameter and kept fixed for the remaining splits.

For the proposed method, we have four balance parameters: λ, γ, μ, and α. We vary the parameters of the projection matrix (μ and α) within the set $\{0.0001, 0.001, 0.01, 0.1, 2, 5, 10, 100, 1000\}$, and the label smoothness term parameter (λ) within $\{0.0001, 0.001, 0.01, 0.1, 2, 5, 10, 20\}$. For γ, we adopt the value proposed by Nie et al. [17].

10.4.3 Data Visualization

In this section, we turn our attention to data visualization to gain insights into how the data are distributed after applying the proposed method.

In our first set of experiments, we utilize the well-known t-SNE algorithm on the PIE and PF01 databases. We apply the t-SNE algorithm on the label matrix calculated using (10.22) and visualize the output for the first 20 classes. Figure 10.1a and b depict the t-SNE output for the PIE database with 5 and 15 labels per class, respectively. Similarly, Fig. 10.2a and b display the t-SNE output for the PF01 database with 3 and 5 labels per class, respectively. As observed, the separation between the classes increases as the number of labeled samples increases, which is also reflected in the classification accuracy.

These visualizations provide a qualitative assessment of the effectiveness of the proposed method in capturing the underlying structure of the data and separating different classes.

In the second experiment, we turn our attention to visualizing the affinity matrices. For this purpose, we select the PF01 database and plot three affinity

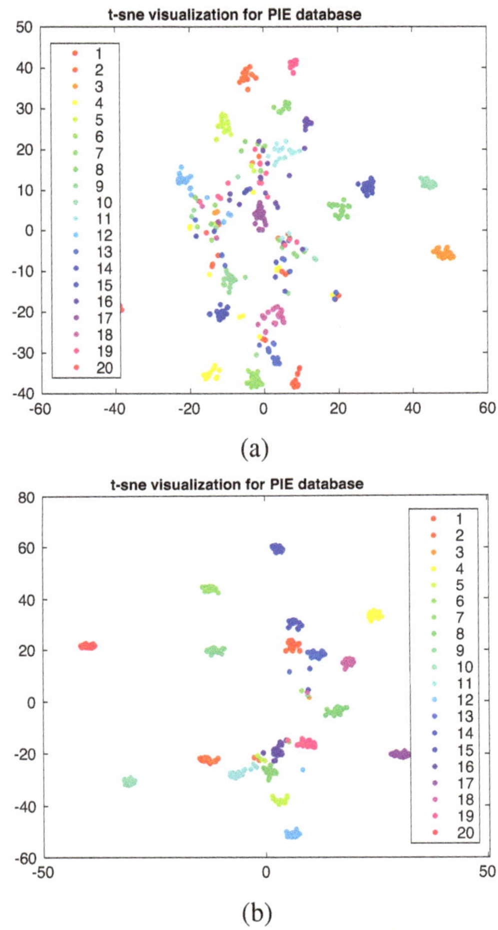

Fig. 10.1 t-SNE visualization for PIE database. (**a**) 5 Labeled per class and (**b**) 15 labeled per class

matrices: the initial graph (Fig. 10.3a), the final graph (Fig. 10.3b), and the ideal graph (Fig. 10.3c) for the first 100 samples.

The initial graph represents the starting point of the algorithm, whereas the final graph illustrates the affinity matrix calculated after the last iteration of the proposed method. On the other hand, the ideal graph represents the desired graph where links are artificially created only for samples within the same class, with no links between samples from different classes.

Upon examination, it is evident that after the optimization process, the final graph closely resembles the ideal graph compared to the initial graph. This indicates that during the optimization process, more accurate links are computed, with a notable enhancement in the connections between samples belonging to the same class. This observation suggests that the proposed method effectively improves the affinity matrix, which is expected to positively impact the overall accuracy of the classification results.

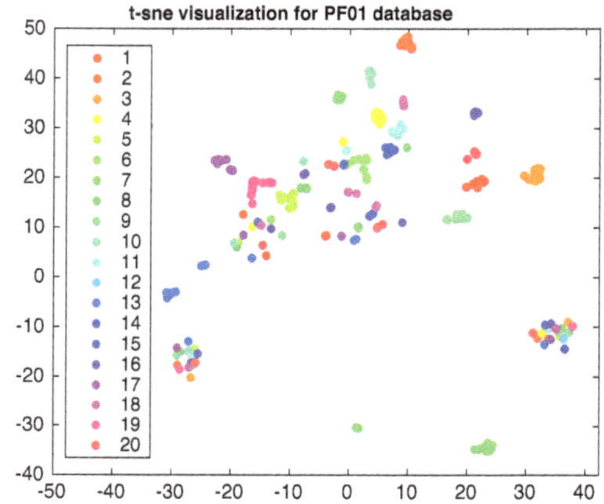

(a)

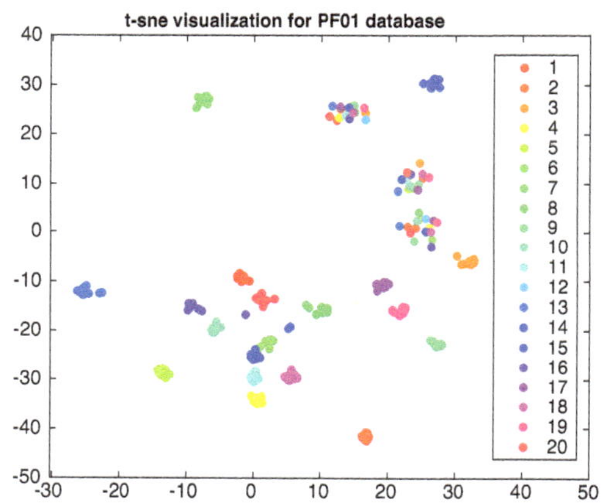

(b)

Fig. 10.2 t-SNE visualization for PF01 database. (**a**) 3 Labeled per class and (**b**) 5 labeled per class

10.4.4 Small Databases

In the case of smaller face databases like PIE, PF01, and FERET, where the number of samples is relatively limited, we adopt a strategy where the test set is not separate, and all samples are utilized for training. From this training data pool, we randomly select l samples as labeled, while the remaining samples are treated as unlabeled.

For these smaller databases, which contain fewer samples, this strategy ensures effective utilization of the available data. The labels for the unlabeled data are inferred using the estimated matrix, $\mathbf{F}$, which is the solution of Eq. (10.15). We compute the label estimate over unlabeled samples across ten different combinations of labeled and unlabeled data splits. The accuracy results for FME with a single feature, FME with feature concatenation, and the proposed methods JGC-MFME and JGC-MFME Weighted are summarized in Table 10.2.

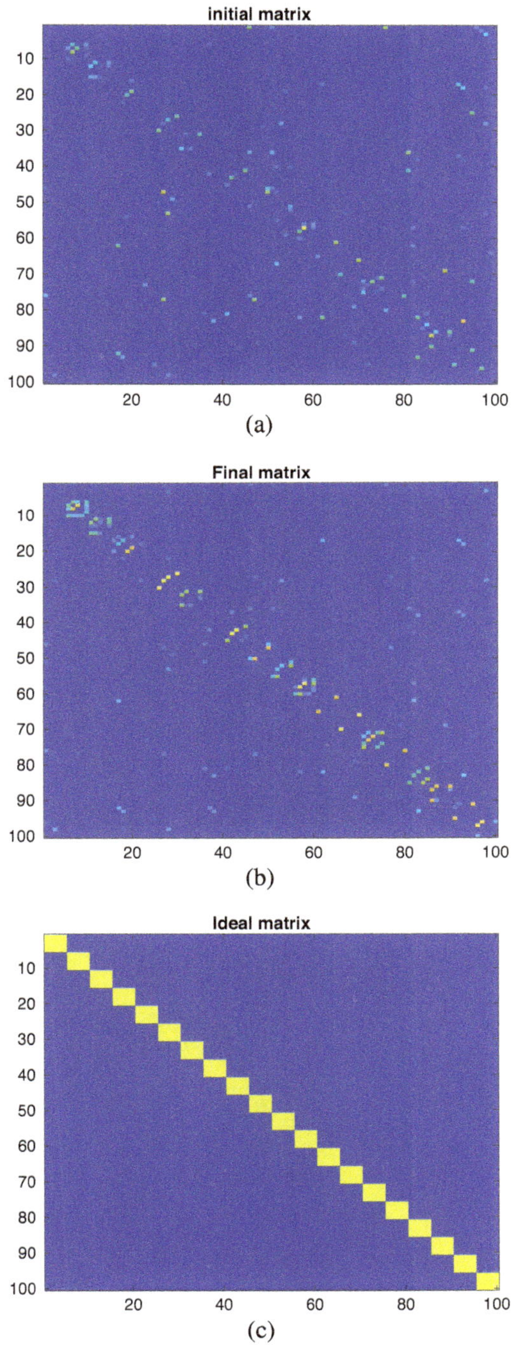

Fig. 10.3 Visualization of affinity matrices. (**a**) The initial graph, (**b**) the final graph, and (**c**) the ideal graph

Table 10.2 Results obtained by the two proposed methods and FME with a single feature. Concatenation of the features is also treated as a single feature. The table shows the average and standard deviation of accuracy obtained on ten splits

PIE						
#Lab./class	Cov	Gabor	LBP	Concat.	JGC-MFME	JGC-MFME weighted
15	80.93±3.71	91.79±3.32	87.75±3.90	94.31±2.88	**94.44±2.76**	94.39±2.86
17	81.93±4.06	91.09±3.08	88.11±2.99	94.54±2.51	94.75±2.58	**94.84±2.47**
FERET						
#Lab./class	Cov	Gabor	LBP	Concat.	JGC-MFME	JGC-MFME weighted
3	59.17±9.53	79.80±8.96	72.01±10.3	83.98±8.06	**85.50±7.64**	84.66±7.74
5	67.27±18.67	86.00±15.38	85.15±17.66	90.72±12.98	**91.77±12.98**	91.45±12.95
PF01						
#Lab. /class	Cov	Gabor	LBP	Concat.	JGC-MFME	JGC-MFME weighted
5	70.24±3.34	84.26±3.76	79.57±2.55	91.30±2.03	92.28±1.61	**92.61±1.74**
7	71.82±5.93	85.81±4.93	80.71±4.51	93.22±2.39	**94.13±2.01**	93.7±2.40

From the table, it is evident that the two proposed methods significantly improves performance compared to using a single feature. Moreover, feature concatenation, as performed in the FME solution, also enhances accuracy compared to single-feature methods. However, the proposed algorithm outperforms both single-feature and concatenation methods, exhibiting higher average accuracy and lower standard deviation. This underscores the effectiveness of the proposed method in leveraging multiple features to better discern discriminant information from the data. The second proposed method can further improve the performance for the PIE and PF01 datasets.

10.4.5 Large Databases

In real-world scenarios, it is impractical to utilize all samples for training, especially in the case of large databases where memory and computational constraints become significant. To address this, we partition the MNIST database into train and test sets. The Train set comprises both labeled and unlabeled data, with the latter participating in model estimation, while the test set remains unseen during training.

In the MNIST database, we select 1000 images from each class for the train set, leaving the remaining images for the test set. Similar to small databases, a subset of l images is labeled, while the rest are considered unlabeled. We compute the accuracy for propagating labels to both unlabeled and test samples across ten different labeled/unlabeled data splits, reporting average and standard deviation accuracy in Table 10.3.

The highest accuracy among the methods is denoted in bold font. We compare the proposed methods JGC-MFME and JGC-MFFM weighted with these algorithms: SNF [27], SMGI [12], DGFLP [15], MLGC [2], AMGL [18], and MMCL [9].

For unlabeled samples, the proposed fusion technique exhibits superior accuracy compared to single-feature methods, highlighting the efficacy of incorporating data smoothness. Additionally, our method outperforms competing fusion techniques due to its integrated learning model, which facilitates better label estimation, and the enhanced performance resulting from fusing features from multiple views.

Furthermore, Table 10.3b reports accuracy for test samples. Notably, transductive methods like SMGI, DGFLP, MLGC, and AMGL cannot process unseen samples; hence, no accuracy is reported for them. However, our proposed method utilizes Eq. (10.7) to estimate labels for unseen samples.

The superior accuracy obtained by our method for both unlabeled and test samples indicates its capability to effectively estimate the underlying data structure and mitigate overfitting or underfitting issues commonly encountered in training set models.

Table 10.3 Results on the MNIST database for different combinations of labeled and unlabeled samples with three different numbers of labeled samples. (Top) Accuracy on unlabeled data estimated by the algorithms. (Bottom) Accuracy on test data estimated via the linear projection of the FME algorithm

MNIST (accuracy on unlabeled data)			
Lab./Class			
Method	20	30	40
Single feature (VGG16)	94.53 ± 0.3	95.17 ± 0.3	95.59 ± 0.3
Single feature (VGG19)	94.35 ± 0.5	94.99 ± 0.5	95.35 ± 0.4
Feature concatenation	95.87 ± 0.4	96.28 ± 0.4	**96.66 ± 0.3**
SNF [27]	94.53 ± 0.6	95.81 ± 0.4	96.49 ± 0.3
SMGI [12]	88.40 ± 0.9	89.66 ± 0.7	89.84 ± 0.7
DGFLP [15]	92.14 ± 0.5	92.70 ± 0.5	93.06 ± 0.3
MLGC [2]	88.20 ± 1.4	89.45 ± 1.0	90.27 ± 0.7
AMGL [18]	94.99 ± 0.5	95.39 ± 0.3	95.67 ± 0.3
MMCL [9]	52.45 ± 14.5	49.98 ± 13.9	49.22 ± 13
JGC-MFME (ours)	**95.88 ± 0.4**	**96.38 ± 0.4**	**96.66 ± 0.3**
JGC-MFME weighted (ours)	**95.89 ± 0.4**	**96.38 ± 0.4**	**96.66 ± 0.3**
MNIST (accuracy on test data)			
Lab./Class			
Method	20	30	40
Single feature (VGG16)	94.98 ± 0.5	95.39 ± 0.3	95.81 ± 0.3
Single feature (VGG19)	95.05 ± 0.5	95.64 ± 0.4	95.97 ± 0.4
Feature concatenation	95.47 ± 0.5	**96.79 ± 0.5**	96.21 ± 0.4
SNF [27]	94.79 ± 0.6	95.97 ± 0.3	96.64 ± 0.2
JGC-MFME (ours)	**96.48 ± 0.4**	96.74 ± 0.3	**97.05 ± 0.3**
JGC-MFME weighted (ours)	**96.48 ± 0.4**	96.75 ± 0.3	**97.05 ± 0.3**

10.5 Conclusion

In our research, we introduced a novel semi-supervised learning algorithm that integrates the principle of data smoothness into its optimization framework. This approach, designed for multi-view data analysis, capitalizes on the wealth of information present in diverse feature descriptors. By simultaneously estimating the unified graph, unlabeled sample labels, and the projection matrix, our method offers a comprehensive solution for leveraging both labeled and unlabeled data effectively.

One key advantage of our approach is its versatility in handling large databases without the need for the entire dataset during the training phase. By utilizing the projection matrix to predict labels for unseen samples, our method demonstrates scalability and applicability to real-world datasets with vast amounts of unlabeled data.

Experimental evaluations underscore the superiority of our proposed method over conventional single-feature approaches and existing fusion techniques. The incorporation of multiple views significantly enhances classification accuracy, showcasing the robustness and effectiveness of our approach in capturing complex data patterns.

Despite its successes, our method has its limitations. For instance, the uniform weighting assigned to different views may not fully exploit the varying informativeness of individual features. To address this, future research could explore adaptive weighting schemes that dynamically adjust the importance of each view based on its contribution to the learning task. The use of a weighted model could improve the performance of semi-supervised classification.

Overall, our study contributes to the advancement of semi-supervised learning techniques, offering a promising avenue for extracting valuable insights from multi-view datasets and facilitating more accurate and reliable data analysis in various domains.

Acknowledgments This work is partially supported by grant GIU23/022 unded by the University of the Basque Country (UPV/EHU), and grant PID2021-126701OB-I00, funded by the Ministerio de Ciencia, Innovaci6n y Universidades, AEI, MCIN/AEI/10.13039/501100011033, and by "ERDF A way of making Europe."

References

1. Ahonen, T., Hadid, A., Pietikainen, M.: Face description with local binary patterns: application to face recognition. IEEE Trans. Pattern Anal. Mach. Intell. **28**(12), 2037–2041 (2006). https://doi.org/10.1109/TPAMI.2006.244
2. An, L., Chen, X., Yang, S.: Multi-graph feature level fusion for person re-identification. Neurocomputing **259**, 39–45 (2017). Multimodal Media Data Understanding and Analytics
3. Angelou, M., Solachidis, V., Vretos, N., Daras, P.: Graph-based multimodal fusion with metric learning for multimodal classification. Pattern Recogn. **95**, 296–307 (2019)
4. Asian Face Image Database PF 01 Intelligent Multimedia. https://api.semanticscholar.org/CorpusID:16663200
5. Bahrami, S., Bosaghzadeh, A., Dornaika, F.: Multi similarity metric fusion in graph-based semi-supervised learning. Computation **7**(1) (2019). https://doi.org/10.3390/computation7010015
6. Bahrami, S., Dornaika, F., Bosaghzadeh, A.: Joint auto-weighted graph fusion and scalable semi-supervised learning. Inform. Fusion **66**, 213–228 (2021)
7. Dornaika, F., Bosaghzadeh, A., Raducanu, B.: Efficient graph construction for label propagation based multi-observation face recognition. In: Salah, A.A., Hung, H., Aran, O., Gunes, H. (eds.) Human Behavior Understanding, pp. 124–135. Springer, Cham (2013)
8. Dornaika, F., Dahbi, R., Bosaghzadeh, A., Ruichek, Y.: Efficient dynamic graph construction for inductive semi-supervised learning. Neural Netw. **94**, 192–203 (2017). https://doi.org/10.1016/j.neunet.2017.07.006
9. Gong, C., Tao, D., Maybank, S.J., Liu, W., Kang, G., Yang, J.: Multi-modal curriculum learning for semi-supervised image classification. IEEE Trans. Image Process. **25**(7), 3249–3260 (2016). https://doi.org/10.1109/TIP.2016.2563981
10. He, F., Nie, F., Wang, R., Hu, H., Jia, W., Li, X.: Fast semi-supervised learning with optimal bipartite graph. IEEE Trans. Knowl. Data Eng. **33**(9), 3245–3257 (2020)
11. Kang, Z., Shi, G., Huang, S., Chen, W., Pu, X., Zhou, J.T., Xu, Z.: Multi-graph fusion for multi-view spectral clustering. Knowl.-Based Syst. **189**, 105102 (2020)

12. Karasuyama, M., Mamitsuka, H.: Multiple graph label propagation by sparse integration. IEEE Trans. Neural Netw. Learn. Syst. **24**(12), 1999–2012 (2013). https://doi.org/10.1109/TNNLS.2013.2271327
13. LeCun, Y., Cortes, C.: MNIST handwritten digit database (2010). http://yann.lecun.com/exdb/mnist/
14. Li, L., He, H.: Bipartite graph based multi-view clustering. IEEE Trans. Knowl. Data Eng. **34**(7), 3111–3125 (2020)
15. Lin, G., Liao, K., Sun, B., Chen, Y., Zhao, F.: Dynamic graph fusion label propagation for semi-supervised multi-modality classification. Pattern Recogn. **68**, 14 – 23 (2017). https://doi.org/10.1016/j.patcog.2017.03.014.
16. Manna, S., Khonglah, J.R., Mukherjee, A., Saha, G.: Robust kernelized graph-based learning. Pattern Recogn. **110**, 107628 (2021)
17. Nie, F., Cai, G., Li, X.: Multi-view clustering and semi-supervised classification with adaptive neighbours. In: Thirty-First AAAI Conference on Artificial Intelligence (2017). https://aaai.org/ocs/index.php/AAAI/AAAI17/paper/view/14833
18. Nie, F., Li, J., Li, X.: Parameter-free auto-weighted multiple graph learning: a framework for multiview clustering and semi-supervised classification. In: Proceedings of the Twenty-Fifth International Joint Conference on Artificial Intelligence, IJCAI'16, pp. 1881–1887. AAAI Press (2016). http://dl.acm.org/citation.cfm?id=3060832.3060884
19. Nie, F., Tian, L., Wang, R., Li, X.: Multiview semi-supervised learning model for image classification. IEEE Trans. Knowl. Data Eng. **32**(12), 2389–2400 (2019)
20. Nie, F., Wang, Z., Wang, R., Li, X.: Adaptive local embedding learning for semi-supervised dimensionality reduction. IEEE Trans. Knowl. Data Eng. **34**(10), 4609–4621 (2021)
21. Nie, F., Xu, D., Tsang, I.W., Zhang, C.: Flexible manifold embedding: a framework for semi-supervised and unsupervised dimension reduction. IEEE Trans. Image Process. **19**(7), 1921–1932 (2010). https://doi.org/10.1109/TIP.2010.2044958
22. Phillips, P.J., Moon, H., Rizvi, S., J. Rauss, P.: The feret evaluation methodology for face-recognition algorithms. IEEE Trans. Pattern Anal. Mach. Intell. **22**, 1090 – 1104 (2000). https://doi.org/10.1109/34.879790
23. Shen, L., Bai, L.: A review on gabor wavelets for face recognition. Pattern Anal. Appl. **9**(2–3), 273–292 (2006)
24. Sim, T., Baker, S., Bsat, M.: The CMU pose, illumination, and expression (pie) database, pp. 46–51 (2002). https://doi.org/10.1109/AFGR.2002.1004130
25. Simonyan, K., Zisserman, A.: Very deep convolutional networks for large-scale image recognition. CoRR abs/1409.1556 (2014). http://arxiv.org/abs/1409.1556
26. Tuzel, O., Porikli, F., Meer, P.: Region covariance: a fast descriptor for detection and classification. In: Leonardis, A., Bischof, H., Pinz, A. (eds.) Computer Vision – ECCV 2006, pp. 589–600. Springer, Heidelberg (2006)
27. Wang, B., Mezlini, A.M., Demir, F., Fiume, M., Tu, Z., Brudno, M., Haibe-Kains, B., Goldenberg, A.: Similarity network fusion for aggregating data types on a genomic scale. Nat. Methods **11**, 333–337 (2014)
28. Wang, H., Yang, Y., Liu, B., Fujita, H.: A study of graph-based system for multi-view clustering. Knowl.-Based Syst. **163**, 1009–1019 (2019)
29. Zheng, F., Liu, Z., Chen, Y., An, J., Zhang, Y.: A novel adaptive multi-view non-negative graph semi-supervised elm. IEEE Access **8**, 116350–116362 (2020)
30. Zhou, D., Bousquet, O., Lal, T.N., Weston, J., Schölkopf, B.: Learning with local and global consistency. In: Thrun, S., Saul, L.K., Schölkopf, B. (eds.) Advances in Neural Information Processing Systems 16, pp. 321–328. MIT Press, Cambridge (2004). http://papers.nips.cc/paper/2506-learning-with-local-and-global-consistency.pdf
31. Zhu, X., Ghahramani, Z., Lafferty, J.D.: Semi-supervised learning using Gaussian fields and harmonic functions. In: Proceedings of the 20th International conference on Machine learning (ICML-03), pp. 912–919 (2003)

Chapter 11
Advancements in Fuzzy Clustering Algorithms for Image Processing: A Comprehensive Review and Future Directions

Vatsala Anand, Deepika Koundal, Thongchai Surinwarangkoon, and Kittikhun Meethongjan

Abstract Fuzzy clustering algorithms have emerged as powerful tools for various image processing tasks, owing to their ability to handle uncertainties and ambiguities inherent in image data. This chapter provides a comprehensive review of recent advancements in fuzzy clustering algorithms for image processing, focusing on applications such as image classification, texture analysis, segmentation of remote sensing images, and object recognition. Specifically, we discuss the principles and applications of fuzzy clustering in image classification, texture analysis, and segmentation tasks, highlighting the advantages and limitations of popular algorithms such as fuzzy C-means (FCM), spatial fuzzy C-means (SFCM), and intuitionistic fuzzy C-means (IFCM). Furthermore, we present a comparative analysis of these algorithms based on their performance metrics and suitability for different image processing tasks. Finally, we identify open challenges and propose potential future

---

V. Anand
Chitkara University Institute of Engineering and Technology, Chitkara University, Rajpura, Punjab, India
e-mail: vatsala.anand@chitkara.edu.in

D. Koundal (✉)
University of Eastern Finland, Kuopio, Finland

School of Computer Science, University of Petroleum & Energy Studies, Dehradun, India

Ho Chi Minh City Open University, Ho Chi Minh City, Vietnam
e-mail: dkoundal@uef.fi; dkoundal@ddn.upes.ac.in; deepika.koundal@ou.edu.vn

T. Surinwarangkoon
Department of Computer Business, College of Innovation and Management, Suan Sunandha Rajabhat University, Bangkok, Thailand
e-mail: thongchai.su@ssru.ac.th

K. Meethongjan
Department of Apply Science, Faculty of Science and Technology, Suan Sunandha Rajabhat University, Bangkok, Thailand
e-mail: kittikhun.me@ssru.ac.th

F. Dornaika et al. (eds.), *Advances in Data Clustering*,
https://doi.org/10.1007/978-981-97-7679-5_11

research directions in fuzzy clustering for image processing, including handling high-dimensional data, integration with deep learning techniques, scalability, interpretability, and addressing complex image structures.

Keywords Fuzzy clustering · Image processing · Image segmentation · Texture analysis · Fuzzy C-Means (FCM) · Information and communication technology · Scientific research · Technology Spatial Fuzzy C-Means (SFCM) · Research development · Intuitionistic Fuzzy C-Means (IFCM) · Comparative analysis · Future directions

11.1 Introduction

Fuzzy clustering method permits information focuses to have a place to numerous clusters at the same time. Unlike conventional clustering strategies, which relegate each information point to a single cluster, fuzzy clustering allots each information point participation esteem for each cluster, demonstrating the degree to which it has a place in that cluster [1]. These participation values are between 0 and 1, with higher values demonstrating more grounded membership.

The key idea behind fuzzy clustering is to represent the uncertainty in data point assignments, allowing for more nuanced and flexible clustering results. By incorporating this uncertainty, fuzzy clustering can handle situations where data points may exhibit characteristics of multiple clusters or where there is an overlap between clusters. This makes fuzzy clustering particularly useful in cases where crisp boundaries between clusters are not well defined [2, 3]. In image processing, fuzzy clustering plays a significant role in image segmentation. Unlike traditional segmentation methods, fuzzy clustering allows for partial memberships of pixels to different clusters, enabling more precise delineation of boundaries between image regions. This flexibility is particularly beneficial in cases where the boundaries are ambiguous or when there is overlapping information between regions.

Additionally, fuzzy clustering offers noise robustness, as it can effectively handle noisy pixels by assigning them partial memberships to clusters based on their similarity to other pixels. This noise resilience leads to more accurate segmentation results, especially in images with high levels of noise.

Moreover, fuzzy clustering is instrumental in feature extraction from images, where it helps in identifying and grouping pixels with similar features, such as color, texture, or intensity. This ability is crucial for tasks like object recognition, content-based image retrieval, and image compression, where capturing meaningful patterns in the image is essential [4].

Fuzzy clustering offers several advantages over traditional clustering techniques. Fuzzy clustering permits information focuses to have a place to numerous clusters at the same time, giving a more adaptable and nuanced representation of the information. This adaptability is especially valuable in cases where information focuses show characteristics of different clusters or when there is cover between clusters. Fuzzy clustering is more robust to noise compared to traditional clustering techniques. By allowing data points to have partial memberships to clusters, fuzzy clustering can reduce the impact of noisy data on the clustering result [5]. Fuzzy clustering can produce smoother and more accurate boundaries between clusters compared to traditional clustering methods, which often result in crisp, discontinuous boundaries. The membership values assigned to data points in fuzzy clustering provide additional information about the degree of association between data points and clusters, which can be valuable for further analysis [6].

However, fuzzy clustering also has some limitations. Fuzzy clustering can be computationally more expensive than traditional clustering techniques, especially for large datasets, due to the need to calculate membership values for each data point. Fuzzy clustering algorithms often require the tuning of parameters, such as the fuzziness parameter, which can affect the clustering result and may require domain knowledge or experimentation to set appropriately [7]. The fuzzy nature of the clustering result can make it more challenging to interpret compared to traditional clustering, where each data point is assigned to a single cluster.

Overall, while fuzzy clustering offers several advantages in terms of flexibility and robustness, its applicability and effectiveness depend on the specific characteristics of the dataset and the clustering task at hand.

The existing applications of fuzzy clustering in image processing include image segmentation, image enhancement, and image classification, texture analysis, and medical image segmentation. Its ability to handle uncertainties and partial memberships makes it a powerful tool for analyzing complex image data.

The rest of the chapter is organized as follows: Sect. 11.2 gives the fuzzy clustering algorithms, in Sect. 11.3 applications in image segmentation are given, Sect. 11.4 shows comparative analysis and future developments, and Sect. 11.5 shows the conclusion.

11.2 Fuzzy Clustering Algorithms

The different fuzzy clustering algorithms are given below:

Fuzzy C-means (FCM) Clustering could be a strategy that permits each information point to have a place for different clusters with changing degrees of membership. Given a dataset

$$X = x_1, x_2, \ldots, x_n$$

and a specified number of clusters k, FCM aims to find k-cluster centroids as $v_1, v_2, \ldots, v_k$ and membership values μ_{ij} for each data point x_i and cluster j. The objective function is to minimize the fuzzy c-means as shown in Eq. (11.1):

$$J = \sum_{i=1}^{n} \sum_{j=1}^{k} \mu_{ij}^{m} ||x_i - v_i||^2 \tag{11.1}$$

where m could be a weighting type that decides the fuzziness of the clustering. The calculation iteratively overhauls the centroids and participation values until joining, allotting information focuses to clusters based on their membership values.

Fuzzy C-means (FCM) clustering works by iteratively allotting data focuses to clusters based on their closeness to cluster centroids and upgrading the centroids based on the mean of the information focuses assigned to each cluster. The calculation initializes cluster centroids randomly and calculates participation values for each information point indicating its degree of having a place to each cluster [8]. These enrollment values are updated in each emphasis based on the separations between information focuses and centroids, weighted by a fluffiness parameter m. The calculation converges when the centroids and enrollment values not alter essentially. FCM is viable for information that will have a place to numerous clusters at the same time.

Fuzzy C-means (FCM) clustering has several key parameters that influence the clustering results. The number of clusters (k) is used to partition the data. Choosing an appropriate value for k is crucial, as it directly impacts the granularity of the clustering. The fuzziness parameter (m) controls the degree of fuzziness in the clustering, with higher values leading to fuzzier clusters. A value of $m = 1$ indicates crisp clustering, while larger values introduce more fuzziness. The criterion used to determine when the algorithm should stop iterating. Common criteria include a maximum number of iterations or a threshold for the change in cluster centroids. The choice of distance metric (e.g., Euclidean, Manhattan, etc.) affects how the similarity between data points and cluster centroids is measured, impacting the clustering results.

The fuzzy C-means (FCM) clustering includes various advantages: FCM allows data points to belong to multiple clusters simultaneously, providing a more flexible representation of complex data structures. FCM is more robust to noisy data compared to traditional clustering methods, as it assigns partial memberships to data points. FCM provides soft clustering, which is useful when data points may belong to multiple clusters or when cluster boundaries are ambiguous. The fuzziness parameter in FCM allows the degree of cluster fuzziness to be adjusted based on the specific requirements of the task.

The fuzzy C-means (FCM) clustering includes various disadvantages. FCM can be computationally expensive, especially for large datasets, due to the iterative nature of the algorithm. FCM is sensitive to the initial guess of cluster centroids, which can lead to different clustering results for different initializations. The soft clustering nature of FCM can make it challenging to interpret the final clustering results, especially when dealing with high-dimensional data.

FCM is suitable for image segmentation tasks where soft boundaries between regions are desired, as they can provide more nuanced segmentations compared to traditional methods. FCM can be used for feature extraction in images by clustering pixels based on their feature vectors, allowing for the identification of meaningful patterns or structures in the image. FCM's robustness to noise makes it suitable for tasks where noise reduction is important, such as denoising images or identifying patterns in noisy data.

Spatial Fuzzy C Means Spatial fuzzy clustering extends traditional fuzzy clustering methods by incorporating spatial information into the clustering process. It considers not only the similarity of data points in feature space but also their spatial proximity [9]. The key working principle involves defining a spatial constraint or penalty term in the clustering objective function, which encourages neighboring data points to belong to the same cluster. This spatial constraint can be based on distance metrics or spatial relationships between data points. By integrating spatial information, spatial fuzzy clustering can produce more coherent and spatially meaningful clusters, particularly in applications such as image segmentation and spatial data analysis.

Spatial fuzzy clustering extends the fuzzy C-means (FCM) algorithm by incorporating spatial information. The objective function includes a spatial constraint term, which penalizes the dissimilarity between neighboring data points. The spatial constraint term can be formulated in Eq. (11.2):

$$\sum_{i=1}^{n}\sum_{j=1}^{k}\mu_{ij}^{m}||x_i - v_i||^2 + \lambda\sum_{i=1}^{n}\sum_{j=1}^{k}w_{ij}d_{ij}||v_i - v_j||^2 \tag{11.2}$$

where w_{ij} is a spatial weight indicating the spatial relationship between data points x_i and x_j, d_{ij} is the spatial distance between data point x_i and x_j, and λ is the parameter balancing the influence of the spatial constraint. The algorithm iteratively updates the cluster centroids and membership values to minimize this objective function, resulting in spatially coherent clusters.

Spatial fuzzy C-means (SFCM) clustering extends fuzzy C-means (FCM) by incorporating spatial information into the clustering process. The key parameters in SFCM and their influence on clustering results are number of clusters (k), fuzziness parameter (m), spatial neighborhood size, spatial influence parameter (λ), distance metric, and stopping criterion [10].

The spatial fuzzy C-means (SFCM) clustering involves various advantages. SFCM considers spatial relationships between data points, resulting in clusters that are based not only on feature similarity but also on spatial proximity. By including spatial information, SFCM produces clusters with better spatial coherence, making it suitable for tasks where spatially contiguous regions are important, such as image segmentation. SFCM's consideration of spatial information can make it more robust to noise, as spatially close points are more likely to be assigned to the same cluster, reducing the impact of noisy outliers. Like traditional fuzzy C-means, SFCM offers soft clustering, allowing data points to belong to multiple clusters with varying degrees of membership, providing a more flexible representation of complex data distributions.

The spatial fuzzy C-means (SFCM) clustering involves various disadvantages. Incorporating spatial information increases the computational complexity of SFCM, especially for large datasets or when using a large spatial neighborhood size. SFCM's performance is influenced by the choice of spatial neighborhood size and spatial influence parameter, which need to be carefully selected based on the specific characteristics of the data and the clustering task. SFCM, like FCM, is sensitive to the initial guess of cluster centroids and membership values, which can lead to different clustering results for different initializations.

SFCM is particularly suitable for image segmentation tasks where spatial coherence is important, such as segmenting anatomical structures in medical images or objects in satellite images. SFCM can be used for texture analysis in images by considering both the texture features and the spatial relationships between pixels, allowing for the identification of complex texture patterns. Apart from image processing, SFCM is also useful for clustering spatial data in other domains such as geography, ecology, and geology, where spatial relationships are key factors in data analysis.

Intuitionistic Fuzzy C-Means (IFCM) This clustering extends traditional fuzzy clustering to handle uncertain and hesitant data. It assigns each data point not only a membership degree but also a nonmembership degree, representing the level of hesitancy or uncertainty in its assignment to clusters. The algorithm iteratively updates the membership and nonmembership degrees based on the distance between data points and cluster centroids. Unlike traditional fuzzy clustering, IFCM considers both the similarity of a data point to a cluster and its dissimilarity to other clusters. This allows IFCM to effectively handle data with varying degrees of uncertainty and hesitation, producing more robust clustering results [11].

Intuitionistic fuzzy C-means (IFCM) clustering extends traditional fuzzy clustering to handle uncertainty and hesitation in data. It introduces two membership functions: membership u_{ij} and nonmembership v_{ij} where i indexes data points and j indexes clusters. The objective function is shown in Eq. (11.3):

$$J = \sum_{i=1}^{n} \sum_{j=1}^{c} [\mu_{ij}^{m} + v_{ij}^{m}] d_{ij}^{2} \tag{11.3}$$

where d_{ij} is the distance between data point i and cluster center j, and m is a weighting exponent. The algorithm iteratively updates μ_{ij} and v_{ij} until convergence, assigning each data point to clusters based on the membership and nonmembership values. This formulation allows IFCM to handle uncertain and hesitant data effectively.

Intuitionistic fuzzy C-means (IFCM) clustering extends traditional fuzzy C-means (FCM) by introducing additional parameters to handle uncertainty and hesitation in data. The key parameters in IFCM and their influence on clustering results are number of clusters (k), fuzziness parameter (m), hesitancy parameter (l), and threshold parameter (epsilon).

IFCM can handle uncertainty and hesitation in data, allowing data points to have varying degrees of confidence in their cluster assignments. This is particularly useful in image processing tasks where the boundaries between objects are ambiguous. Like traditional fuzzy C-means (FCM), IFCM provides soft clustering, allowing data points to belong to multiple clusters simultaneously with varying degrees of membership. This flexibility can capture complex data distributions more effectively. IFCM's ability to handle uncertainty makes it more robust to noisy data compared to crisp clustering methods, as data points can express their uncertainty about their cluster assignments. IFCM is suitable for clustering complex datasets where data points may exhibit uncertainty or hesitation in their cluster memberships.

IFCM can be computationally expensive, especially for large datasets, due to the iterative nature of the algorithm and the need to calculate membership and hesitancy values for each data point. IFCM's performance can be sensitive to the choice of parameters such as the fuzziness parameter, hesitancy parameter, and threshold parameter, which need to be carefully selected based on the characteristics of the data and the clustering task. IFCM is well suited for image segmentation tasks where the boundaries between objects are not well defined, as they can handle uncertainty and hesitation in data point assignments. IFCM can also be used for feature extraction in images by clustering pixels based on their feature vectors, allowing for the identification of meaningful patterns or structures in the image. IFCM's ability to handle uncertainty makes it suitable for texture analysis tasks in images, where textures may exhibit varying degrees of complexity and ambiguity.

Spatial Intuitionistic Fuzzy C-Means (SIFCM) Clustering combines the principles of intuitionistic fuzzy C-means (IFCM) clustering with spatial information. The mathematical formulation of SIFCM includes additional terms to incorporate spatial constraints. The objective function of SIFCM can be defined in Eq. (11.4):

$$J = \sum_{i=1}^{n}\sum_{j=1}^{c}[\mu_{ij}^{m} + v_{ij}^{m}]d_{ij}^{2} + \lambda\sum_{i=1}^{n}\sum_{j=1}^{k} w_{ij}||v_i - v_j||^2 \tag{11.4}$$

The algorithm iteratively updates the membership and nonmembership degrees, cluster centroids, and spatial weights until convergence, resulting in spatially coherent clusters with uncertainty handling capabilities [12].

The working principles of SIFCM can be summarized as first initializing cluster centroids, membership degrees, nonmembership degrees, and spatial weights. Update the membership degrees μ_{ij} and nonmembership degrees v_{ij} for each data point i and cluster j. Update the cluster centroids v_j using the updated membership and nonmembership degrees. Update the spatial weights w_{ij} based on the spatial relationships between data points i and j. Iterate steps second and forth until convergence criteria are met (e.g., small changes in centroids, membership, and nonmembership values). Assign each data point to the cluster with the highest membership degree and use the nonmembership degree as a measure of uncertainty. By incorporating spatial information and handling uncertainty, SIFCM produces clustering results that are more suitable for image processing tasks where spatial coherence and uncertainty management are crucial. The key parameters in SIFCM and their influence on clustering results are number of clusters (k), fuzziness parameter (m), hesitancy parameter (l), spatial neighborhood size, spatial influence parameter (λ), distance metric, and stopping criterion.

The advantages of spatial intuitionistic fuzzy C-means (SIFCM) clustering is given as SIFCM produces clusters that are not only based on feature similarity but also on spatial proximity, leading to more spatially coherent clusters compared to traditional fuzzy clustering methods. SIFCM can handle uncertainty and hesitation in data, allowing data points to have varying degrees of confidence in their cluster assignments. This is particularly useful in image processing tasks where the boundaries between objects are ambiguous. Like traditional fuzzy C-means (FCM) and intuitionistic fuzzy C-means (IFCM), SIFCM provides soft clustering, allowing data points to belong to multiple clusters simultaneously with varying degrees of membership. SIFCM's ability to handle uncertainty makes it more robust to noisy data compared to crisp clustering methods, as data points can express their uncertainty about their cluster assignments.

The disadvantages of spatial intuitionistic fuzzy C-means (SIFCM) clustering are given as incorporating spatial information increases the computational complexity of SIFCM, especially for large datasets or when using a large spatial neighborhood size. SIFCM's performance is influenced by the choice of parameters such as the fuzziness parameter, hesitancy parameter, and spatial influence parameter, which need to be carefully selected based on the characteristics of the data and the clustering task.

SIFCM is well suited for image segmentation tasks, suitable for texture analysis tasks in images. The mathematical formulation of neutrosophic clustering is based on neutrosophic sets, which are characterized by three components: truth membership T, indeterminacy I, and falsity membership (F) as shown in Eq. (11.5).

$$J = \sum_{i=1}^{n} \sum_{j=1}^{k} \left[\mu_{ij}^{T} D\left((x_i, c_j)^2 \right) + \mu_{ij}^{I} I\left((x_i, c_j)^2 \right) + \mu_{ij}^{F} F\left((x_i, c_j)^2 \right) \right] \tag{11.5}$$

$D(x_i, c_j)$ is the distance between data point x_i and cluster center c_j, $I(x_i, c_j)$ is the indeterminacy between x_i and c_j, $F(x_i, c_j)$ is the falsity between x_i and c_j, and μ_{ij}^{T}, μ_{ij}^{I}, μ_{ij}^{F} are the truth membership, indeterminancy, and falsity membership degrees.

The optimization process involves iteratively updating the cluster centers and the membership degrees until convergence. Neutrosophic clustering aims to find clusters that not only minimize the distance between data points and cluster centers but also account for the indeterminacy and falsity of the data, making it suitable for handling uncertain and inconsistent data. Neutrosophic clustering algorithms represent data points as neutrosophic sets, which consist of three components: a membership function (T), a nonmembership function (I), and an indeterminacy function (F). These functions indicate the degree of truth, falsity, and indeterminacy of a data point belonging to a cluster.

The objective function of a neutrosophic clustering algorithm is formulated to minimize the sum of the distances between data points and cluster centroids, taking into account the neutrosophic components. The objective function aims to minimize the membership and nonmembership values while maximizing the indeterminacy values, leading to clusters that are well separated, yet allow for uncertainty.

Neutrosophic clustering algorithms iteratively update the cluster centroids and the neutrosophic components of data points. The updated rules adjust the membership, nonmembership, and indeterminacy values based on the distances between data points and cluster centroids, as well as the influence of neighboring data points.

Neutrosophic clustering algorithms handle indeterminacy by allowing data points to have varying degrees of indeterminacy in their cluster assignments. This flexibility allows the algorithms to capture the uncertainty and ambiguity inherent in real-world data.

Neutrosophic clustering algorithms converge when there is no significant change in the cluster centroids and the neutrosophic components of data points between iterations. The convergence criterion ensures that the clustering result is stable and representative of the underlying data distribution. Overall, neutrosophic clustering algorithms aim to provide a flexible and robust framework for clustering data with uncertain or indeterminate characteristics, making them suitable for various applications in image processing, pattern recognition, and data mining.

Some of the key parameters and their influence on neutrosophic clustering algorithms are number of clusters (k), threshold parameter (ϵ), fuzziness parameter (m), distance metric, and stopping criterion. Choosing appropriate values for these parameters is essential for obtaining meaningful and useful clustering results with neutrosophic clustering algorithms, as they can significantly impact the clustering outcome.

The advantages of neutrosophic clustering algorithms include the following: neutrosophic clustering algorithms can handle uncertainty, indeterminacy, and ambiguity in data, making them suitable for tasks where the data is not clearly defined or boundaries between clusters are ambiguous. Neutrosophic clustering algorithms provide soft clustering, allowing data points to belong to multiple clusters simultaneously with varying degrees of membership and nonmembership. Neutrosophic clustering algorithms are robust to noise, as they can accommodate data points that do not clearly belong to any cluster. Representation of partial knowledge: neutrosophic clustering algorithms can represent partial knowledge about the data, allowing for more nuanced and flexible clustering results.

The disadvantages of neutrosophic clustering algorithms include the following: Neutrosophic clustering algorithms can be computationally expensive, especially for large datasets, due to the need to calculate membership, nonmembership, and indeterminacy degrees for each data point. The performance of neutrosophic clustering algorithms is sensitive to the choice of parameters such as the fuzziness parameter, threshold parameter, and initialization method, which need to be carefully selected based on the characteristics of the data and the clustering task.

Neutrosophic clustering algorithms are well suited for image segmentation tasks and also can be used for texture analysis in images, where textures may exhibit varying degrees of complexity and ambiguity. The soft clustering capability allows for the identification of complex texture patterns.

Spatial Neutrosophic Clustering Algorithms Extend neutrosophic clustering to incorporate spatial information into the clustering process. The objective function of spatial neutrosophic clustering algorithms can be formulated as shown in Eq. (11.6):

$$J = \sum_{i=1}^{n} \sum_{j=1}^{c} w_{ij}[(1-\alpha)D(x_i, v_i) + \alpha S(x_i, v_i)]u_{ij} + \lambda R(U) \tag{11.6}$$

where w_{ij} is the spatial weight, $D(x_i, v_i)$ is the measure of data similarity, $S(x_i, v_i)$ is the measure of spatial coherence, $R(U)$ is the regularization term, α balances the influence of data similarity, and λ controls the influence of regularization term. The working principles of spatial neutrosophic clustering algorithms can be summarized as follows:

The algorithm starts by initializing the cluster centroids and the neutrosophic membership values for each data point. Each data point is assigned to one or more clusters based on its similarity to the cluster centroids and its spatial relationships with other data points. The neutrosophic membership values uij are updated to reflect the degree to which each data point belongs to each cluster. After the assignment step, the cluster centroids are updated based on the current assignment of data points. The spatial relationships between data points are also considered in this step to ensure spatial coherence. Steps 2 and 3 are repeated iteratively until a stopping criterion is met, such as a maximum number of iterations or a small change in the cluster centroids. The spatial relationships between data points are incorporated into the clustering process using spatial weights. These weights reflect the spatial proximity between data points and are used to calculate the spatial coherence term in the objective function. Neutrosophic membership values allow for uncertainty in the assignment of data points to clusters. Data points can have varying degrees of truth-membership, indeterminacy, and falsehood-membership to each cluster, capturing the uncertainty in the clustering process. By integrating spatial information and uncertainty handling into the clustering process, spatial neutrosophic clustering algorithms can produce clustering results that are well suited for image processing tasks where spatial coherence and uncertainty are important.

Spatial neutrosophic clustering algorithms have several key parameters that influence their clustering results. These parameters include the number of clusters (k), fuzziness parameter (m), spatial neighborhood size, spatial influence parameter (λ), regularization parameter (λ), and threshold parameter (ϵ). Choosing appropriate values for these parameters is crucial for obtaining meaningful and useful clustering results with spatial neutrosophic clustering algorithms, as they can significantly impact the clustering outcome [12].

The advantages of spatial neutrosophic clustering algorithms include spatial neutrosophic clustering algorithms consider spatial relationships between data points, leading to spatially coherent clustering results. This is particularly useful in image processing tasks where spatial context is important, such as image segmentation. Neutrosophic clustering algorithms can handle uncertainty, indeterminacy, and ambiguity in data, making them suitable for tasks where the data is not clearly defined or boundaries between clusters are ambiguous. Like traditional fuzzy C-means (FCM) and intuitionistic fuzzy C-means (IFCM), neutrosophic clustering algorithms provide soft clustering, allowing data points to belong to multiple clusters simultaneously with varying degrees of membership and nonmembership. Neutrosophic clustering algorithms are robust to noise, as they can accommodate data points that do not clearly belong to any cluster. Neutrosophic clustering algorithms offer flexibility in representing partial knowledge about the data, allowing for more nuanced and flexible clustering results.

The disadvantages of spatial neutrosophic clustering algorithms are as follows: Incorporating spatial information increases the computational complexity of neutrosophic clustering algorithms, especially for large datasets or when using a large spatial neighborhood size. The performance of neutrosophic clustering algorithms is sensitive to the choice of parameters such as the fuzziness parameter, threshold parameter, and spatial influence parameter, which need to be carefully selected based on the characteristics of the data and the clustering task.

Spatial neutrosophic clustering algorithms are well suited for image segmentation tasks where the boundaries between objects are not well defined and there is uncertainty in the data, as they can handle ambiguity and indeterminacy in data point assignments while considering spatial relationships. Neutrosophic clustering algorithms can be used for texture analysis in images, where textures may exhibit varying degrees of complexity and ambiguity. The incorporation of spatial information also helps in capturing spatial texture patterns.

11.3 Applications in Image Segmentation

Fuzzy clustering has been widely used in image segmentation due to its ability to handle uncertainties and ambiguities in image data. Fuzzy clustering algorithms assign each pixel in an image a membership value indicating the likelihood of belonging to different clusters, allowing for soft segmentation where pixels can belong to multiple clusters simultaneously. This approach contrasts with traditional hard segmentation methods, which assign each pixel to a single cluster.

One of the key advantages of fuzzy clustering for image segmentation is its ability to capture the gradual transitions and uncertainties present in natural images. This is particularly useful in cases where pixel intensities or features do not cleanly delineate object boundaries. By allowing pixels to have partial memberships in multiple clusters, fuzzy clustering can produce more accurate and visually appealing segmentations compared to hard clustering methods.

Several fuzzy clustering algorithms have been applied to image segmentation, with fuzzy C-means (FCM) being one of the most popular. FCM minimizes the difference between pixel values and cluster centroids weighted by the membership values, effectively clustering pixels based on their similarity to cluster centroids. Extensions of FCM, such as spatially constrained FCM (SFCM), incorporate spatial information into the clustering process, promoting spatial coherence in the segmented regions.

Another variant, Possibilistic C-Means (PCM), is more robust to noise and outliers compared to FCM. PCM replaces the membership values in FCM with possibility values, which measure the degree to which a pixel can be reasonably assigned to a cluster. This approach reduces the sensitivity of the clustering algorithm to noisy data, improving segmentation results, especially in noisy images.

In recent years, researchers have explored hybrid approaches that combine fuzzy clustering with other techniques to improve segmentation accuracy. For example, combining fuzzy clustering with region merging algorithms can refine the segmentation results by merging regions based on both pixel intensities and spatial coherence. Similarly, combining fuzzy clustering with deep learning techniques has shown promise in capturing complex image features for segmentation tasks.

Despite its advantages, fuzzy clustering for image segmentation has some limitations. One challenge is the computational complexity, especially for large datasets or high-dimensional feature spaces. Additionally, the choice of clustering parameters, such as the number of clusters and the fuzziness parameter, can significantly impact the segmentation results and may require manual tuning.

In conclusion, fuzzy clustering offers a flexible and powerful approach to image segmentation, allowing for soft segmentation and handling uncertainties inherent in image data. While it has some limitations, ongoing research aims to address these challenges and further improve the effectiveness of fuzzy clustering for image segmentation tasks.

Fuzzy clustering-based segmentation algorithms are evaluated using various metrics to assess their performance and compare them to ground truth or reference segmentations. These metrics provide quantitative measures of segmentation quality, including accuracy, completeness, and consistency. Some commonly used evaluation metrics for fuzzy clusteringbased segmentation algorithms include:

Dice Similarity Coefficient (DSC): DSC measures the overlap between the segmented regions and the ground truth regions.

Jaccard Index (JI): JI is similar to DSC but emphasizes the intersection over the union of the segmented and ground truth regions.

Sensitivity and Specificity: Sensitivity (true positive rate) measures the proportion of true positive pixels correctly classified as belonging to the segmented region, while specificity (true negative rate) measures the proportion of true negative pixels correctly classified as not belonging to the segmented region.

Precision and Recall: Precision measures the proportion of true positive pixels among all pixels classified as belonging to the segmented region, while recall measures the proportion of true positive pixels among all pixels that actually belong to the ground truth region.

Fowlkes-Mallows Index (FMI): FMI is a measure of similarity between two clustering results, considering both the number of pairs of points that are correctly or incorrectly clustered together. It ranges from 0 to 1, with higher values indicating better clustering results.

Adjusted Rand Index (ARI): ARI measures the similarity between two clustering results, accounting for chance agreement. It adjusts the Rand Index to account for the expected similarity between two random clustering results. ARI ranges from -1 to 1, with 0 indicating random clustering and 1 indicating identical clustering.

Normalized Mutual Information (NMI): NMI measures the amount of information shared between two clustering results, normalized by the total information in each clustering. It ranges from 0 to 1, with higher values indicating more similar clustering results.

These evaluation metrics provide quantitative measures of the performance of fuzzy clustering-based segmentation algorithms, helping researchers and practitioners assess their effectiveness and compare them to other segmentation approaches.

11.4 Comparative Analysis and Future Directions

A comparative analysis of these algorithms based on their performance in different image processing tasks is given below:

Algorithm	Advantage	Disadvantage
Fuzzy C-means (FCM)	FCM is computationally efficient and easy to implement. It provides soft clustering, allowing pixels to belong to multiple clusters simultaneously. FCM is widely used in image segmentation tasks	FCM is sensitive to noise and outliers, which can affect its clustering performance. It also requires the selection of a fuzziness parameter, which can impact the quality of the clustering results
Possibilistic C-means (PCM)	PCM is more robust to noise and outliers compared to FCM. It replaces the membership values in FCM with possibility values, which reduces the impact of noisy data on clustering results	PCM can be computationally more expensive than FCM due to the additional calculations required for possibility values. It also requires the selection of a fuzziness parameter, similar to FCM
Spatial fuzzy C-means (SFCM)	SFCM incorporates spatial information into the clustering process, leading to more spatially coherent clustering results. It is particularly useful in image segmentation tasks where spatial context is important	SFCM can be computationally more expensive than FCM or PCM due to the additional calculations required for spatial information. It also requires the selection of spatial parameters, which can affect the quality of the clustering results
Hybrid fuzzy clustering Algorithms	Hybrid fuzzy clustering algorithms combine fuzzy clustering with other techniques, such as genetic algorithms or neural networks, to improve clustering performance. These algorithms can capture complex patterns in the data and produce more accurate clustering results	Hybrid fuzzy clustering algorithms can be more complex and computationally expensive than traditional fuzzy clustering algorithms. They also require the selection of additional parameters, which can impact their performance

In summary, fuzzy clustering algorithms offer a flexible and powerful approach to various image processing tasks. The choice of algorithm depends on the specific requirements of the task, such as the presence of noise, the need for spatial coherence, and the complexity of the data. Researchers and practitioners can select the most suitable fuzzy clustering algorithm based on these requirements to achieve optimal results in their image processing tasks.

Open challenges and potential future research directions in fuzzy clustering for image processing include:

Handling High-Dimensional Data: Fuzzy clustering algorithms may struggle with high-dimensional data, such as hyperspectral images, due to increased computational complexity and difficulty in defining meaningful distances between data points. Future research could focus on developing efficient algorithms or dimensionality reduction techniques tailored to high-dimensional data.

Integration with Deep Learning: Integrating fuzzy clustering with deep learning techniques, such as convolutional neural networks (CNNs), could lead to more powerful and flexible image segmentation methods. Future research could explore hybrid approaches that combine the strengths of both fuzzy clustering and deep learning for improved segmentation accuracy. Scalability: Fuzzy clustering algorithms may face scalability issues when applied to large-scale image processing tasks. Future research could focus on developing parallel or distributed algorithms that can handle large datasets efficiently.

Interpretability and Explainability: Fuzzy clustering algorithms often lack interpretability, making it challenging to understand and validate the segmentation results. Future research could focus on developing more interpretable models or methods for explaining the clustering decisions.

Handling Complex Image Structures: Fuzzy clustering algorithms may struggle with segmenting images containing complex structures or textures. Future research could focus on developing algorithms that can better capture and represent the complex relationships within such images.

Addressing these challenges and exploring these research directions could lead to significant advancements in fuzzy clustering for image processing, enabling more accurate and robust segmentation of various types of images.

Future research directions in fuzzy clustering for image processing could include:

Dynamic Fuzzification: Developing adaptive fuzzification strategies that dynamically adjust the fuzziness parameter based on the characteristics of the data could improve the performance of fuzzy clustering algorithms, especially in handling complex and heterogeneous image data.

Incorporating Domain Knowledge: Integrating domain-specific knowledge into fuzzy clustering algorithms could enhance their performance and interpretability. This could involve incorporating constraints or prior knowledge about the image structures or relationships between image regions into the clustering process.

Sparse Representation: Exploring sparse representation techniques within the context of fuzzy clustering could lead to more efficient and compact representations of image data, improving both computational efficiency and clustering quality.

Deep Fuzzy Clustering: Investigating the integration of fuzzy clustering with deep learning approaches, such as autoencoders or generative adversarial networks, could lead to more powerful and flexible clustering models for image processing tasks.

Evaluation Metrics: Developing new evaluation metrics or extending existing ones to better assess the performance of fuzzy clustering algorithms, especially in comparison to ground truth or reference segmentations, could provide more reliable benchmarks for algorithm evaluation.

Deep learning techniques, such as convolutional neural networks (CNNs), have shown remarkable success in various image processing tasks, including image segmentation and clustering. By integrating deep learning with fuzzy clustering, several benefits can be achieved:

Feature Learning: Deep learning models can automatically learn hierarchical features from raw image data, capturing complex patterns and structures that may be challenging to define manually. These learned features can then be used as input to fuzzy clustering algorithms, improving the quality of clustering results.

Improved Robustness: Deep learning models are often more robust to noise and variations in image data compared to traditional handcrafted features. By incorporating deep learning features into fuzzy clustering, the clustering algorithm can better handle noisy or complex image data, leading to more accurate segmentation results.

Hybrid Architectures: Hybrid approaches that combine the strengths of deep learning and fuzzy clustering can leverage the flexibility of fuzzy clustering in handling uncertainties and soft assignments, while also benefiting from the feature learning capabilities of deep learning models. This can lead to more flexible and adaptive clustering algorithms that can better capture the underlying structure of image data.

Semantic Segmentation: Deep learning models, particularly fully convolutional networks (FCNs), have been successful in semantic segmentation tasks, where each pixel is assigned a class label. By integrating fuzzy clustering with FCNs, it is possible to combine the semantic segmentation capabilities of FCNs with the soft clustering nature of fuzzy clustering, leading to more accurate and semantically meaningful image segmentation.

Overall, the potential of deep learning and hybrid approaches for enhancing the performance of fuzzy clustering in image processing is significant. By leveraging the strengths of both approaches, it is possible to develop more robust, adaptive, and accurate clustering algorithms for a wide range of image processing tasks.

11.5 Conclusions

Fuzzy clustering has the potential to advance the field of image processing in several ways:

Handling Uncertainty: Fuzzy clustering can effectively model uncertainty and ambiguity in image data, allowing for more flexible and nuanced segmentation compared to traditional hard clustering methods. This ability is particularly useful in applications where precise boundaries between objects are difficult to define.

Soft Segmentation: Fuzzy clustering provides soft segmentation, where each pixel is assigned a membership value indicating its likelihood of belonging to different clusters. This can lead to more natural and visually appealing segmentations, especially in areas with gradual transitions or overlapping structures.

Integration with Other Techniques: Fuzzy clustering can be easily integrated with other image processing techniques, such as feature extraction, dimensionality reduction, and classification. This flexibility allows for the development of customized and adaptive image processing pipelines tailored to specific applications.

Robustness to Noise: Fuzzy clustering algorithms are often more robust to noise and outliers compared to traditional clustering methods. This robustness is particularly valuable in image processing tasks where the input data may be corrupted or contain artifacts.

Semantic Understanding: By providing a framework for modeling uncertainty and soft assignments, fuzzy clustering can contribute to the development of image processing algorithms that have a deeper understanding of the semantic content of images. This can lead to advancements in tasks such as object recognition, scene understanding, and image retrieval.

Overall, the potential of fuzzy clustering in advancing the field of image processing lies in its ability to handle uncertainty, provide soft segmentation, integrate with other techniques, and contribute to semantic understanding. By leveraging these capabilities, fuzzy clustering can help address complex image processing challenges and open up new avenues for research and application development.

Future research in fuzzy clustering for image processing could focus on several areas to further enhance its capabilities. First, developing more efficient and scalable algorithms to handle large-scale image datasets could improve the applicability of fuzzy clustering in real-world applications. Second, exploring novel hybrid approaches that combine fuzzy clustering with other machine learning techniques, such as deep learning or reinforcement learning, could lead to more robust and adaptive clustering algorithms. Lastly, investigating the use of fuzzy clustering in emerging applications, such as medical imaging, remote sensing, and video processing, could uncover new opportunities for advancing the field of image processing.

References

1. Xu, S., Hao, Z., Zhu, Y., Wang, Z., Xiao, Y., Liu, B.: Semi-supervised fuzzy clustering algorithm based on prior membership degree matrix with expert preference. Expert Syst. Appl. **238**, 121812 (2024)
2. Yu, H., Jiang, L., Fan, J., Xie, S., Lan, R.: A feature-weighted suppressed possibilistic fuzzy c-means clustering algorithm and its application on color image segmentation. Expert Syst. Appl. **241**, 122270 (2024)

3. Kmita, K., Kaczmarek-Majer, K., Hryniewicz, O.: Explainable impact of partial supervision in semi-supervised fuzzy clustering. IEEE Trans. Fuzzy Syst. (2024)
4. Zhang, C., Chen, L., Shi, Z., Ding, W.: Latent information-guided one-step multi-view fuzzy clustering based on cross-view anchor graph. Inform. Fusion **102**, 102025 (2024)
5. Liu, F., Dong, T., Liu, Q., Liu, Y., Li, S.: Combining fuzzy clustering and improved long short-term memory neural networks for short-term load forecasting. Electric Power Syst. Res. **226**, 109967 (2024)
6. Kaushal, M., Danish Lohani, Q.M., Castillo, O.: Weighted intuitionistic fuzzy c-means clustering algorithms. Int. J. Fuzzy Syst. **26**(3), 943–977 (2024)
7. Ma, Q., Zhu, X., Zhao, X., Zhao, B., Fu, G., Zhang, R.: An equidistance index intuitionistic fuzzy c-means clustering algorithm based on local density and membership degree boundary. Appl. Intell. **54**(4), 3205–3221 (2024)
8. Xing, H., Zhang, M., Tong, Q., Zeng, X., Chen, H.: An improved interval type-2 possibilistic c-means clustering algorithm for interclass maximization. J. Intell. Fuzzy Syst.. Preprint. (2024)
9. Bhattacharya, T., Soares, G.A.B.E., Chopra, H., Rahman, M.M., Hasan, Z., Swain, S.S., Cavalu, S.: Applications of phyto-nanotechnology for the treatment of neurodegenerative disorders. Materials **15**(3), 804 (2022)
10. Sharma, B., Koundal, D.: Cattle health monitoring system using wireless sensor network: a survey from innovation perspective. IET Wirel. Sensor Syst. **8**(4), 143–151 (2018)
11. Ramalingam, S., Dhanasekaran, S., Sinnasamy, S.S., Salau, A.O., Alagarsamy, M.: Performance enhancement of efficient clustering and routing protocol for wireless sensor networks using improved elephant herd optimization algorithm. Wirel. Netw. **30**(3), 1773–1789 (2024)
12. Gupta, M., Mishra, A.: A systematic review of deep learning based image segmentation to detect polyp. Artif. Intell. Rev. **57**(1), 1–53 (2024)

www.ingramcontent.com/pod-product-compliance
Lightning Source LLC
Chambersburg PA
CBHW072011120125
20267CB00001B/22
9789819776788